Basic Math Refresher

"I would like to thank my father for teaching me to appreciate numbers and graphs."

—Dr. Stephen Hearne

Basic Math Refresher

Stephen Hearne, Ph.D.

Research & Education Association
Visit our website at
www.rea.com

Research & Education Association
61 Ethel Road West
Piscataway, New Jersey 08854
E-mail: info@rea.com

BASIC MATH REFRESHER

Published 2010

Printed in the United States of America

Library of Congress Control Number 2004115677

ISBN-13: 978-0-7386-0052-9
ISBN-10: 0-7386-0052-0

REA® is a registered trademark of
Research & Education Association, Inc.

ABOUT OUR AUTHOR

Stephen Hearne is currently a Professor at Skyline College in San Bruno, California where he teaches Quantitative Reasoning among other subjects.

For the past twenty years, **Stephen Hearne** has worked as a mathematics tutor teaching students of all ages in math, algebra, statistics, and test preparation.

He prides himself in being able to make the complex simple.

ACKNOWLEDGMENTS

In addition to our author, REA would like to thank **Jeremy Rech**, Graphic Designer, for designing the book and typesetting the manuscript; **Molly Solanki**, Associate Editor, for coordinating revisions; **Rachel DiMatteo**, Graphic Designer, for typesetting revisions; **Larry B. Kling**, Vice President, Editorial, for supervising development; and **Pam Weston**, Vice President, Publishing, for ensuring press readiness.

We also extend thanks to **Adel Arshaghi** for his editorial contributions.

ABOUT RESEARCH & EDUCATION ASSOCIATION

Founded in 1959, Research & Education Association is dedicated to publishing the finest and most effective educational materials—including software, study guides, and test preps—for students in middle school, high school, college, graduate school, and beyond.

Table of Contents

Preface

It takes a long time to learn something, but it takes only a short time to forget it. The consolation is that you can learn the material faster and more easily the second time around. Some of the information remains in your memory. You just need to be reminded.

This book was written for people who took basic math in high school, but have forgotten some of the operations and want to brush-up on their math skills. It is intended to help those who are preparing to take a math test, as well as parents who want to help their children with their math homework. This refresher book can sharpen your skills if you are applying for a job, or have recently received a job, that requires math.

The goal of this book is to give you a concise and understandable review of basic math so that you can use it in your everyday life. Math is useful because it allows you to describe things and to make decisions based on numbers. The calculations in this book are meant to be carried out with a simple hand calculator, although they can be done with paper and pencil.

—Stephen Hearne, Ph.D.

Chapter 1

Addition

Addition is the adding of two or more numbers together.
The result is called the sum. Here are some examples.

EXAMPLE SET 1

34	49	88	831
+ 95	+ 50	+ 69	+ 583
129	99	157	1,414

36	71	22	134
+ 628	+ 126	+ 254	+ 361
664	197	276	495

EXAMPLE SET 2

```
  511        404      5,836         85
+ 147    + 2,830    + 5,358    + 7,149
  658      3,234     11,194      7,234


  209         57        530     72,249
+ 5,409   + 1,911    + 7,469   + 41,692
5,618      1,968      7,999    113,941
```

 EXAMPLE SET 3

```
  4,037     261,885      6,822      84,299
+ 40,581  + 548,917   + 948,624  + 7,825,734
 44,618     810,802    955,446    7,910,033


    192      50,796    1,688,378       2,193
+ 93,050   + 748,129  + 2,777,800  + 3,409,801
 93,242     798,925   4,466,178    3,411,994
```

EXAMPLE SET 4

28	405	1,178
58	750	3,994
45	201	5,329
69	623	2,780
+ 97	+ 973	+ 6,303
297	2,952	19,584

EXAMPLE SET 5

150	47	40,400
229	890	5,855
579	38	38,449
48	244	2,134
530	7	72,749
9	66	638
42	670	6,574
29	7,196	73
949	8	3,862
+ 100	+ 86	+ 11,071
2,665	9,252	181,805

EXAMPLE WORD PROBLEM 1

A family wants to go out to the movie theater. Movie tickets sell for $9.50 for adults and $4.75 for children. Popcorn is $5.00. Candy costs $2.25 a box. How much will it cost for one adult with two children to go to the movies if they buy two popcorns and one box of candy?

Solution

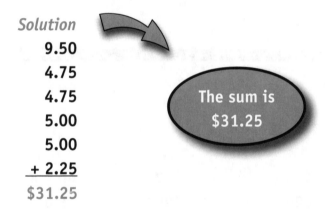

9.50
4.75
4.75
5.00
5.00
+ 2.25
$31.25

The sum is $31.25

 EXAMPLE WORD PROBLEM 2

A carpenter wants to install crown wood molding around the perimeter of a window that is 48 inches tall and 108

inches long. How many total inches of molding does the carpenter need?

Solution

```
   48
   48
  108
+ 108
  312
```

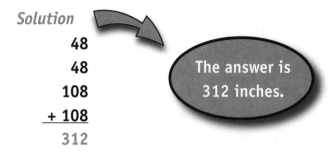

The answer is 312 inches.

  EXAMPLE WORD PROBLEM 3

The library at a major university owns 1,247,606 books. A government grant is allowing the library to purchase 235,000 additional books. Further, a private donation from an alumni member will fund the purchase of 50,500 more books. How many books will the library own all together?

Solution

```
1,247,606
  235,000
+  50,500
1,533,106
```

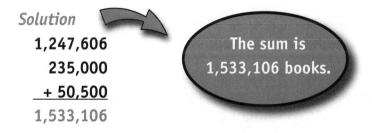

The sum is 1,533,106 books.

EXAMPLE WORD PROBLEM 4

A small farm in the Central Valley grows table grapes. It has been in business for four years. The first year, it produced 30,196 pounds of grapes. The second year, it produced 31,842 pounds of grapes. The third year, it produced 26,128 pounds of grapes. The fourth year, it produced 34,243 pounds of grapes. How many pounds of grapes has this farm produced?

Solution

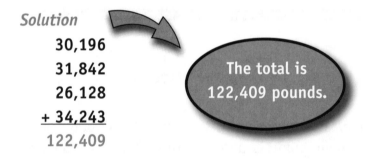

```
   30,196
   31,842
   26,128
 + 34,243
  122,409
```

The total is 122,409 pounds.

 ## EXAMPLE WORD PROBLEM 5

At You-Build-It computer store, you save money on labor because you build your own computer. A central processing unit costs $875. A monitor costs $225. A mouse costs $25.

A mouse pad costs $5. A pair of speakers costs $45. A modem costs $35, and a keyboard costs $40. What is the total cost for a complete computer?

Solution

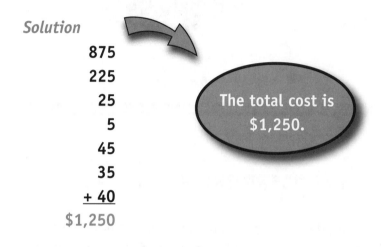

```
    875
    225
     25
      5
     45
     35
  + 40
$1,250
```

The total cost is $1,250.

 EXAMPLE WORD PROBLEM 6

Big Al's Car Dealership has 227 cars on their lot. They expect to buy 60 more cars at an auction next week. They are also waiting on the delivery of 75 cars that the owner, Big Al, purchased on an out of town buying trip. In addition, another 53 cars are expected to arrive shortly from a cross-county dealership. When he gets them all together, how many cars will Big Al have?

Solution

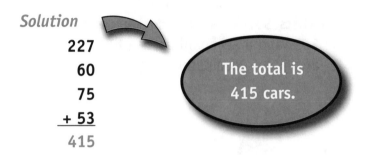

227
60
75
+ 53
415

The total is
415 cars.

 EXAMPLE WORD PROBLEM 7

A real-estate development company wants to build an office building. They buy a piece of property in the downtown area for $2,175,000. A short time later, the company purchases the property next to it for $1,500,000 and the one to next that for $1,950,000. Then, the company paid construction costs of $6,000,000 to have the offices built. What is the total amount of money invested?

Solution

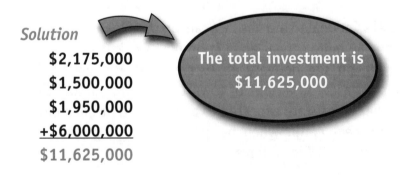

$2,175,000
$1,500,000
$1,950,000
+$6,000,000
$11,625,000

The total investment is
$11,625,000

Practice Problems*

*Answers to Practice Problems on Pages 12-14

1. A camper bought the following camping equipment: a tent for $105, a sleeping bag for $82, a lantern for $36, a camping stove for $48, a tarp for $22, an axe for $27, a flashlight for $9, and some cookware for $15. What was the total cost?

2. A circus show performs five days a week. On Wednesday, the show had 8,632 visitors. On Thursday, it had 4,170 visitors. On Friday, it had 3,937 visitors. On Saturday, it had 5,472 visitors. On Sunday, it had 4,622 visitors. How many visitors did the circus get this week?

3. A five-person family used 14,586 gallons of water in January. They used 12,529 gallons in February, 10,799 in March, and 11,791 in April. How may total gallons of water did this family use over this four-month period?

4. A landscape management company is designing several yards around a home owned by one of its customers. Because of the shape of the lot, they will need to bring in some topsoil. The architect estimates that the back yard will need 1,200 cubic feet of soil. The front yard needs 320 cubic feet of soil. There are also two side yards. One will need 225 cubic feet of soil, while the other will need 200. What is the total number of cubic feet of soil that this project requires?

5. A real-estate broker sold a luxury house for $1,350,000. The new owner has agreed to pay the sales commission, which is $40,500. The mortgage fee is $1,000. In addition, repairs to the property, prior to move in, will cost $26,175. What is the total amount that the new owner must pay?

Calculations*

*Answers to Calculations on Page 14

```
  647          120          471           95
+ 279        + 234        6,135        5,494
                          6,606        5,589
   84          325           634          547
+ 831        + 8,573          28          314
                           3,915       39,981
                         + 9,336       36,956
1,228          185                      76,937
+ 1,947      + 5,527                         72
                                           995
                                        39,411
6,157                                 + 68,076
+ 4,026
```

Answers*

*Answers to the preceding Practice Problems and Calculations

1. Question: A camper bought the following camping equipment: a tent for $105, a sleeping bag for $82, a lantern for $36, a camping stove for $48, a tarp for $22, an axe for $27, a flashlight for $9, and some cookware for $15. What was the total cost?

Answer:
$105 + $82 + $36 + $48 + $22 + $27 + $9 + $15 = $344

2. Question: A circus show performs five days a week. On Wednesday, the show had 8,632 visitors. On Thursday, it had 4,170 visitors. On Friday, it had 3,937 visitors. On Saturday, it had 5,472 visitors. On Sunday, it had 4,622 visitors. How many visitors did the circus get this week?

Answer:
8,632 + 4,170 + 3,937 + 5,472 + 4,622 = 26,833 visitors

3. Question: A five-person family used 14,586 gallons of water in January. They used 12,529 gallons in February, 10,799 in March, and 11,791 in April. How may total gallons of water did this family use over this four-month period?

Answer:
14,586 + 12,529 + 10,799 + 11,791 = 49,705 gallons

4. Question: A landscape management company is designing several yards around a home owned by one of its customers. Because of the shape of the lot, they will need to bring in some topsoil. The architect estimates that the back yard will need 1,200 cubic feet of soil. The front yard needs 320 cubic feet of soil. There are also two side yards. One will need 225 cubic feet of soil, while the other will need 200. What is the total number of cubic feet of soil that this project requires?

Answer:
1,200 + 320 + 225+ 200 = 1,945 cubic feet

5. Question: A real-estate broker sold a luxury house for $1,350,000. The new owner has agreed to pay the sales commission, which is $40,500. The mortgage fee is $1,000. In addition, repairs to the property, prior to move in, will cost $26,175. What is the total amount that the new owner must pay?

▼

Answer:
$1,350,000 + $40,500 + $1,000 + $26,175 = $1,417,675

CALCULATIONS

647	120	471	95
+ 279	+ 234	6,135	5,494
926	354	6,606	5,589
		634	547
84	325	28	314
+ 831	+ 8,573	3,915	39,981
915	8,898	+ 9,336	36,956
		27,125	76,937
1,228	185		72
+ 1,947	+ 5,527		995
3,175	5,712		39,411
			+ 68,076
6,157			274,467
+ 4,026			
10,183			

Subtraction is the taking of a number from another number. Subtraction tells you the difference between two quantities. Here are some examples.

 EXAMPLE SET 1

59	33	894	385
-20	-18	-208	-138
39	15	686	247

514	126	254	629
-45	-71	-22	-349
469	55	232	280

EXAMPLE SET 2

645	3,233	1,785	3,769
-387	- 898	-1,228	-85
258	2,335	557	3,684

6,057	5,222	7,937	95,113
-859	-80	-697	-75,467
5,198	5,142	7,240	19,646

 EXAMPLE SET 3

51,692	372,996	957,624	8,936,845
-5,048	-159,828	-6,823	-95,377
46,644	213,168	950,801	8,841,468

83,060	859,235	2,057,600	3,018,904
-292	-60,876	-1,617,873	-50,400
82,768	798,359	439,727	2,968,504

EXAMPLE SET 4

24,330	860,139	50,307	78,978
-4,735	-91,254	-40,995	-1,294
19,595	768,885	9,312	77,684

36,036	487,343	7,334,880	5,250,500
-27,387	-413,406	-2,595,499	-2,135,000
8,649	73,937	4,739,381	3,115,500

 EXAMPLE WORD PROBLEM 1

The Fruity-Filling Company produces cans of cherry pie filling. Last year, they sold 484,950 cans of cherry pie filling. This year, they sold 563,441 cans. How many more cans did they sell this year compared with last year?

Solution

563,441
-484,950
78,491

The answer is 78,491 cans.

EXAMPLE WORD PROBLEM 2

At Mile-High Hot Air Balloon Rides, they promise you a scenic view. On one particular trip, the pilot took the balloon up to a cruising altitude of 10,000. Then, they made a series of descents. They descended 2,300 feet, then 1,800 feet, then 500 feet, and 975 feet. What was their new altitude?

Solution

```
  2,300
  1,800
    500
+ 975
  5,575  feet drop
```

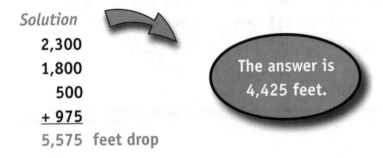

The answer is 4,425 feet.

```
10,000 original altitude
- 5,575
  4,425
```

 EXAMPLE WORD PROBLEM 3

A bride-to-be is planning her wedding. She has $1,000 to spend on flowers, but she has a long list. She needs a bridal bouquet for herself ($150), a boutonniere for the groom ($25), a bouquet for the maid of honor ($50), and a boutonniere for the best man ($25). She also needs corsages for her mother and the mother of the groom ($60), a basket of petals for the flower girl ($25), flowers for the altar ($200), and flowers for the reception ($300). How much money will she have left over?

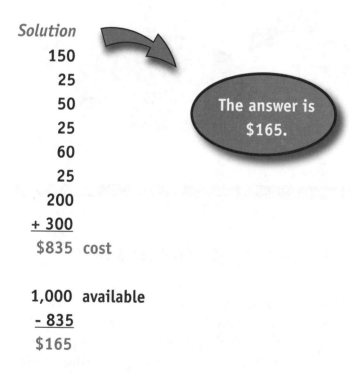

Solution

```
  150
   25
   50
   25
   60
   25
  200
+ 300
$835  cost
```

The answer is $165.

```
1,000  available
- 835
$165
```

EXAMPLE WORD PROBLEM 4

The Department of Fish and Wildlife report that the duck population last year for one western state was 391,953 ducks. This year, the duck population is 533,730. How many ducks is this increase?

Solution

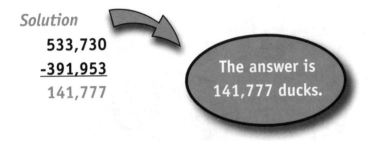

533,730
-391,953
141,777

The answer is 141,777 ducks.

 EXAMPLE WORD PROBLEM 5

A property owner owns a lot on a cliff that overlooks the ocean. When he bought the land, it was an acre big, or 43,560 square feet. The first big storm eroded 2,087 square-feet of land. The second big storm eroded 1,669 square-feet. How many square-feet of land does the owner have left?

Solution

  2,087
+1,669

3,756 square feet eroded

The answer is
39,804 square feet.

43,560 square feet at start
-3,756

39,804

 EXAMPLE WORD PROBLEM 6

There is $50.00 in the food budget. You go to the grocery store to buy fruits and vegetables. You buy green beans for $1.49, potatoes for 39¢, onions for 79¢, bananas for 49¢, tomatoes for 99¢, squash for 79¢, pears for 99¢, and asparagus for $2.59. How much change will you get back from your $50.00?

Solution

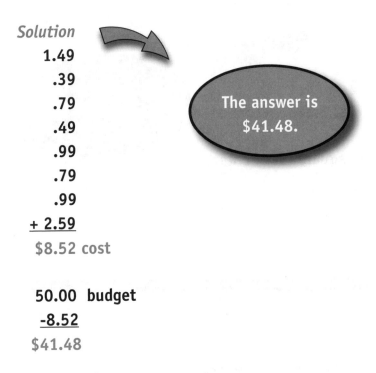

```
   1.49
    .39
    .79
    .49
    .99
    .79
    .99
 + 2.59
 $8.52 cost
```

The answer is $41.48.

```
 50.00  budget
 -8.52
 $41.48
```

 EXAMPLE WORD PROBLEM 7

An executive at a big company is having her office decorated. She has a budget of $20,000 to spend. The items that she wants are: a desk for $5,500, a couch for $3,250, chairs for $2,000, lamps for $2,350, carpet for $2,600, drapes for $1,500, and art for $1,000. How much under budget is she?

Solution

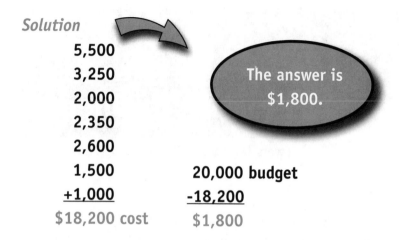

5,500
3,250
2,000
2,350
2,600
1,500
+1,000
$18,200 cost

20,000 budget
-18,200
$1,800

The answer is
$1,800.

 EXAMPLE WORD PROBLEM 8

The water level in a local lake filled up to 17,623 acres in the spring. In the winter, the lake was drained down to 10,517 acres. How big of a change is this?

Solution

17,623
- 10,517
7,106

The answer is
7,106 acres.

Practice Problems*

*Answers to Practice Problems on Pages 27-29

1. The first year that Hope Community College opened, the student enrollment was 8,500. The next year, enrollment fell by 350. The year after that, it fell by 1,090 students, and this year it has 750 fewer students. How many students are enrolled this year?

2. The yearly profits for a big company rose slightly from $14,791,500 last year to $14,808,900 this year. How much of an increase is this?

3. Five years ago, the population of one suburban city was 107,592 people. Today, this city has a population of 123,624 people. How many additional people live in this city compared with five years ago?

4. National Carpet Cleaning has offices nationwide. Their western division serviced 20,565 customers over the last six months. During the same time period, their eastern division serviced 26,783 customers. How many more customers did the eastern division have compared with the western division?

5. A rancher in the southwest owned 31,000 head of cattle. He sold 8,742 of them at a livestock auction in town. Then, he sold 3,065 of his cattle to an east-coast distributor. Lastly, he sold 2,900 of his herd to a neighbor. How many cattle does the rancher have left?

Calculations*

*Answers to Calculations on Page 29

3,428	5,409	1,942	91,221
-973	-14	-967	-24,090

12,288	954,526	575,149	5,516,979
-4,710	-522,283	-9,411	-36,036

91,816	807,489	4,657,479	3,905,543
-71,499	-800,446	-2,494,518	-1,202,746

Answers*

*Answers to the preceding Practice Problems and Calculations

1. Question: The first year that Hope Community College opened, the student enrollment was 8,500. The next year, enrollment fell by 350. The year after that, it fell by 1,090 students, and this year it has 750 fewer students. How many students are enrolled this year?

Answer:
350 + 1,090 + 750 = 2,190
This is the total number of fewer students.

8,500 - 2,190 = 6,310 students

2. Question: The yearly profits for a big company rose slightly from $14,791,500 last year to $14,808,900 this year. How much of an increase is this?

Answer:
14,808,900 - 14,791,500 = $17,400

3. **Question:** Five years ago, the population of one suburban city was 107,592 people. Today, this city has a population of 123,624 people. How many additional people live in this city compared with five years ago?

> *Answer:*
> 123,624 - 107,592 = 16,032 people

4. **Question:** National Carpet Cleaning has offices nationwide. Their western division serviced 20,565 customers over the last six months. During the same time period, their eastern division serviced 26,783 customers. How many more customers did the eastern division have compared with the western division?

> *Answer:*
> 26,783 - 20,565 = 6,218 more customers

5. **Question:** A rancher in the southwest owned 31,000 head of cattle. He sold 8,742 of them at a livestock auction in town. Then, he sold 3,065 of his cattle to an east-coast distributor. Lastly, he sold 2,900 of his herd to a neighbor. How many cattle does the rancher have left?

Answer:
8,742 + 3,065 + 2,900 = 14,707
This is the total number of cattle sold.

31,000 - 14,707 = 16,293 cattle

CALCULATIONS

3,428	5,409	1,942	91,221
-973	-14	-967	-24,090
2,455	5,395	975	67,131

12,288	954,526	575,149	5,516,979
-4,710	-522,283	-9,411	-36,036
7,578	432,243	565,738	5,480,943

91,816	807,489	4,657,479	3,905,543
-71,499	-800,446	-2,494,518	-1,202,746
20,317	7,043	2,162,961	2,702,797

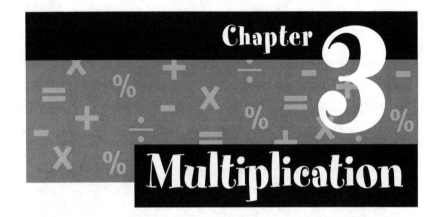

Chapter 3

Multiplication

Multiplication tells you the result obtained by repeating a number a specific amount of times. The result of multiplying two numbers together is called the product. Here are some examples.

 EXAMPLE SET 1

29	22	65	90
x 7	x 9	x 93	x 8
203	198	6,045	720

94	78	6	78
x 65	x 18	x 43	x 19
6,110	1,404	258	1,482

EXAMPLE SET 2

88	979	16	732
x 35	x 63	x 797	x 58
3,080	61,677	12,752	42,456

56	207	35	155
x 100	x 60	x 57	x 884
5,600	12,420	1,995	137,020

 EXAMPLE SET 3

47	270	558	91
x 11	x 41	x 64	x 86
517	11,070	35,712	7,826

134	37	85	715
x 55	x 477	x 632	x 741
7,370	17,649	53,720	529,815

EXAMPLE SET 4

438	772	3,955	460
x 58	x 251	x 725	x 4,753
25,404	193,772	2,867,375	2,186,380

6,691	3,932	1,536	2,924
x 941	x 358	x 30	x 8,613
6,296,231	1,407,656	46,080	25,184,412

 EXAMPLE WORD PROBLEM 1

It takes a skilled craftsman 12 hours of labor to make a pair of custom cowboy boots. How many hours will it take to make 25 pairs of boots?

Solution

12
x 25
300

The answer is 300 hours.

EXAMPLE WORD PROBLEM 2

A start-up company needs to buy 75 new copy machines. Each machine costs $885. How much will the copy machines cost?

Solution

```
    885
   x 75
 $66,375
```

The answer is $66,375.

 EXAMPLE WORD PROBLEM 3

The speed of the space shuttle in low earth orbit is 17,500 miles per hour. At this rate, how far will the space shuttle travel in 66 hours?

Solution

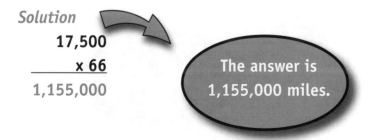

```
   17,500
     x 66
 1,155,000
```

The answer is 1,155,000 miles.

EXAMPLE WORD PROBLEM 4

Pet-O-Rama, a nationwide pet retail company, sold 15,394 pet rabbits to its customers last year. Each rabbit sold for $14.00. How much money did this company take in on rabbit sales?

Solution

15,394
x 14
$215,516

The answer is $215,516.

 EXAMPLE WORD PROBLEM 5

The United States sells 160,000 tons of corn each month to a particular foreign country. After 18 months, how much corn will this country have purchased?

Solution

160,000
x 18
2,880,000

The answer is 2,880,000 tons.

EXAMPLE WORD PROBLEM 6

Ladybugs are good for gardens because they eat aphids. Aphids can harm plants. A ladybug can eat 57 aphids a day. At this rate, how many aphids can 250 ladybugs eat in one day?

Solution

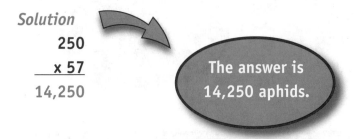

```
    250
  x  57
 14,250
```

The answer is 14,250 aphids.

 ## EXAMPLE WORD PROBLEM 7

A contractor is building a tract of new homes. The total number of windows that need to be built for all of the homes is 658. Each window uses 16 panes of glass. How many panes of glass will the contractor need all together?

Solution

```
    658
  x  16
 10,528
```

The answer is 10,528 panes.

EXAMPLE WORD PROBLEM 8

A big oil well in Texas produces 14,500 barrels of crude oil each day. How many barrels of oil will it produce in one year? (Use 365 days in a year.)

Solution

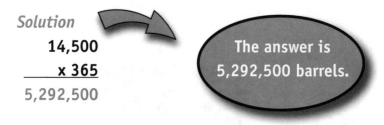

```
   14,500
  x 365
5,292,500
```

The answer is 5,292,500 barrels.

 ## EXAMPLE WORD PROBLEM 9

At Magnify Laboratories, 125 blood samples can be analyzed per day. How many samples can be analyzed in 90 days?

Solution

```
   125
  x 90
11,250
```

The answer is 11,250 samples.

EXAMPLE WORD PROBLEM 10

A workman is laying decorative tile on a kitchen floor. The kitchen is a rectangle with the dimensions 24 feet by 22 feet. How many square feet of tile does the workman need?

Solution

```
   24
 x 22
  528
```

The answer is 528 square feet.

 ## EXAMPLE WORD PROBLEM 11

A child saved 37¢ a day for 3 years. How much money did the child save? (Use 365 days in a year.)

Solution

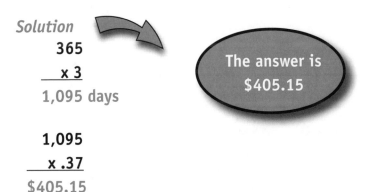

```
  365
  x 3
1,095 days
```

The answer is $405.15

```
  1,095
  x .37
$405.15
```

EXAMPLE WORD PROBLEM 12

Each student at Daisy Elementary School needs 10 pencils to start the year. There are 357 students all together and each pencil costs 4¢. How much will the pencils cost?

Solution

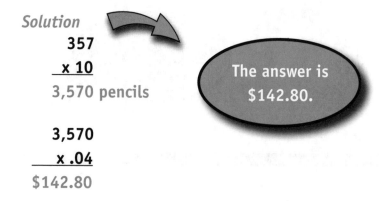

357
x 10
3,570 pencils

The answer is
$142.80.

3,570
x .04
$142.80

Practice Problems*

*Answers to Practice Problems on Pages 42-44

1. An auditorium has 18 rows of chairs with 25 chairs in each row. How many chairs does the auditorium have all together?

2. Each week, 865,400 cars cross over the Golden Gate Bridge. How many cars cross over this bridge in a year? (There are 52 weeks in a year.)

3. The management at On-Time Delivery company wants to buy each of its 1,241 employees a desk clock. Each clock costs $13.57. What is the total cost for this many clocks?

4. A workman is laying carpet in the living room of a large home. The living room is 38 feet long and 25 feet wide. How many square feet of carpet are needed?

5. An industrial painting company recently purchased 215 gallons of paint at $9.45 per gallon. What was the total cost?

6. A freeway commuter drives 44 miles round trip to work each day, 5 days a week for 50 weeks a year. How many miles a year does this commuter drive round trip to work?

Calculations*

*Answers to Calculations on Page 44

30	21	67	83
x 6	x 4	x 68	x 40

384	891	68	912
x 92	x 603	x 359	x 259

3,801	4872	468	3,341
x 37	x 336	x 8,728	x 6,186

Answers*

*Answers to the preceding Practice Problems and Calculations

1. Question: An auditorium has 18 rows of chairs with 25 chairs in each row. How many chairs does the auditorium have all together?

> *Answer:*
> **18 x 25 = 450 chairs**

2. Question: Each week, 865,400 cars cross over the Golden Gate Bridge. How many cars cross over this bridge in a year? (There are 52 weeks in a year.)

> *Answer:*
> **52 x 865,400 = 45,000,800 cars**

3. Question: The management at On-Time delivery company wants to buy each of its 1,241 employees a desk clock. Each clock costs $13.57. What is the total cost for this many clocks?

> *Answer:*
> 1,241 x 13.57 = $16,840.37

4. Question: A workman is laying carpet in the living room of a large home. The living room is 38 feet long and 25 feet wide. How many square feet of carpet are needed?

> *Answer:*
> 38 x 25 = 950 square feet

5. Question: An industrial painting company recently purchased 215 gallons of paint at $9.45 per gallon. What was the total cost?

> *Answer:*
> 215 x 9.45 = $2,031.75

6. Question: A freeway commuter drives 44 miles round trip to work each day, 5 days a week for 50 weeks a year. How many miles a year does this commuter drive round trip to work?

Answer:
44 x 5 = 220 miles per week

220 x 50 = 11,000 miles per year

CALCULATIONS

30	21	67	83
x 6	x 4	x 68	x 40
180	84	4,556	3,320

384	891	68	912
x 92	x 603	x 359	x 259
35,328	537,273	24,412	236,208

3,801	4872	468	3,341
x 37	x 336	x 8,728	x 6,186
140,637	1,636,992	4,084,704	20,667,426

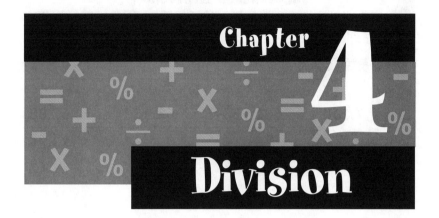

Chapter 4
Division

Division is finding out how many times a number goes into another number. The result is called the quotient. Here are some examples.

 EXAMPLE SET 1

342	115	2,996	803
÷ 9	÷ 23	÷ 4	÷ 11
38	5	749	73

156	340	9,504	1,024
÷ 52	÷ 85	÷ 96	÷ 16
3	4	99	64

EXAMPLE SET 2

7,160	1,958	68,182	6,288
÷ 8	÷ 22	÷ 73	÷ 262
895	89	934	24

4,032	8,879	4,953	12,300
÷ 192	÷ 13	÷ 381	÷ 4
21	683	13	3,075

 EXAMPLE SET 3

36,630	4,393	17,178	71,604
÷ 55	÷ 191	÷ 21	÷ 918
666	23	818	78

59,052	16,262	45,262	9,860
÷ 798	÷ 94	÷ 854	÷ 290
74	173	53	34

EXAMPLE SET 4

199,609	493,554	552,720	7,421,568
÷ 47	÷ 86	÷ 735	÷ 96
4,247	5,739	752	77,308

6,171,025	494,239	7,001,807	4,013,674
÷ 985	÷ 971	÷ 71	÷ 553
6,265	509	98,617	7,258

 EXAMPLE WORD PROBLEM 1

Arizona has a population of 2,736,000 people. This state occupies 114,000 square miles of land. How many people per square mile are there in Arizona?

Solution

2,736,000
÷ 114,000
24

The answer is 24 people per square mile.

EXAMPLE WORD PROBLEM 2

The will of a late real-estate tycoon is being settled in court. In the will, all assets are to be divided evenly between all of his remaining 20 relatives. The estate is worth $4,802,000. How much money will each heir receive?

Solution

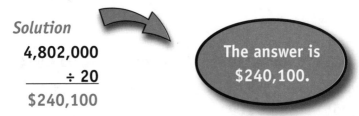

    4,802,000
       ÷ 20
    $240,100

The answer is $240,100.

 ## EXAMPLE WORD PROBLEM 3

At Rising-Star Electric Guitars, workers manufactured 3,750 guitars last year. If they worked 250 days, how many guitars did they build on the average day?

Solution

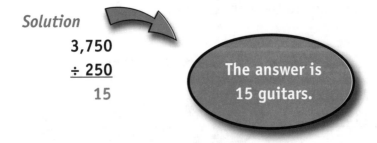

      3,750
      ÷ 250
        15

The answer is 15 guitars.

EXAMPLE WORD PROBLEM 4

A school bond measure is on the ballot for the next city election. The amount of the bond proposal is $6,000,000. If the bond measure passes, each of the city's 40,000 homeowners will have to pay an equal share of the bond. How much money would each homeowner have to pay if the bond measure passes?

Solution

6,000,000
÷ 40,000
$150

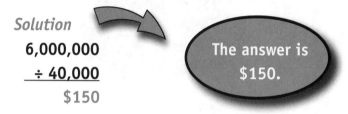

The answer is $150.

 ## EXAMPLE WORD PROBLEM 5

A taxi driver traveled a total of 364 miles in one day. He used 13 gallons of gasoline. How many miles per gallon did his taxi get?

Solution

364
÷13
28

The answer is 28 miles per gallon.

EXAMPLE WORD PROBLEM 6

The Good Neighbors Fence Company sells pre-built, 8-foot fence sections. How many sections will a person need if he wants to fence a yard with a perimeter of 272 feet?

Solution

272
÷ 8
34

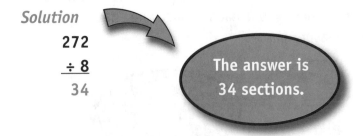

The answer is 34 sections.

 EXAMPLE WORD PROBLEM 7

A passenger train traveled west 2,058 miles from Memphis, Tennessee to Los Angeles, California. The total trip took 49 hours. What was the train's average rate of speed?

Solution

2,058
÷ 49
42

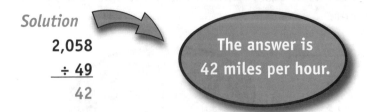

The answer is 42 miles per hour.

EXAMPLE WORD PROBLEM 8

At Turn-Em Real-Estate Company, they are selling an old school building with an asking price of $775,704. The building is 64,642 square feet in size. What is the cost per square foot?

Solution

775,704
÷ 64,642
$12

The answer is $12 per square foot.

 ## EXAMPLE WORD PROBLEM 9

Wide-Open-Spaces Horse Ranch is located on 1,120 acres of lush green land. They own 56 horses. How many acres per horse is this?

Solution

1,120
÷ 56
20

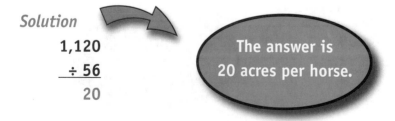

The answer is 20 acres per horse.

EXAMPLE WORD PROBLEM 10

A commercial fishing boat brings in 10,800 pounds of fish. If it is to be divided evenly among the 12 crewmembers, how many pounds will each fisherman get?

Solution

   10,800
   ÷ 12
    900

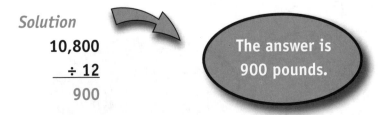

The answer is 900 pounds.

EXAMPLE WORD PROBLEM 11

At O-My Jewelry Store, they are selling diamond engagement rings on credit with no money down and no interest. The ring that the bride-to-be desires costs $8,100. How much would the monthly payments be on a 36-month loan?

Solution

   8,100
   ÷ 36
   $225

The answer is $225 per month.

EXAMPLE WORD PROBLEM 12

A farmer has 10,000 tomato plants that need planting. He plowed 25 rows in a field. How many plants go in each row to make them even?

Solution

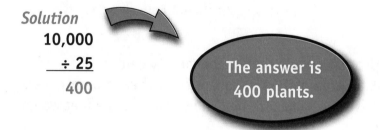

10,000
÷ 25
400

The answer is
400 plants.

*Answers to Practice Problems on Pages 57-59

1. Thirty-six people share a community garden. The garden covers an area of 12,960 square feet. If each person were to receive the same size plot, how many square feet would it be?

2. A worker earns $42,000 a year doing skilled labor. How much does this worker earn each month?

3. A man drove his car on a road trip that was 765 miles long. His average rate of speed was 45 miles per hour. How long did the trip take?

4. The profit sharing division of one company has $1,896,000 in its account. The money is to be divided evenly between its 1,200 employees. How much money will each employee get?

5. New York State has a population of 18,976,410 people. This state covers 47,205 square miles of land. How many people per square mile are there in New York?

6. A five-person family used 12,540 gallons of water during a 30-day period. How many gallons per day did they use?

Calculations*

*Answers to Calculations on Page 59

8,650 ÷ 50	19,564 ÷ 292	5,346 ÷ 22	42,842 ÷ 691
41,710 ÷ 430	54,488 ÷ 98	67,968 ÷ 708	48,000 ÷ 960
352,702 ÷ 59	576,621 ÷ 79	629,760 ÷ 984	8,740,460 ÷ 92

Answers*

*Answers to the preceding Practice Problems and Calculations

1. Question: Thirty-six people share a community garden. The garden covers an area of 12,960 square feet. If each person were to receive the same size plot, how many square feet would it be?

> *Answer:*
> 12,960 ÷ 36 = 360 square feet

2. Question: A worker earns $42,000 a year doing skilled labor. How much does this worker earn each month?

> *Answer:*
> 42,000 ÷ 12 = $3,500 per month

3. Question: A man drove his car on a road trip that was 765 miles long. His average rate of speed was 45 miles per hour. How long did the trip take?

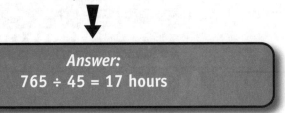

Answer:
765 ÷ 45 = 17 hours

4. Question: The profit sharing division of one company has $1,896,000 in its account. The money is to be divided evenly between its 1,200 employees. How much money will each employee get?

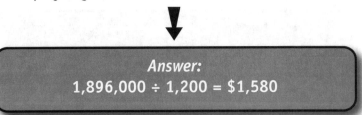

Answer:
1,896,000 ÷ 1,200 = $1,580

5. Question: New York State has a population of 18,976,410 people. This state covers 47,205 square miles of land. How many people per square mile are there in New York?

Answer:
18,976,410 ÷ 47,205 = 402 people per square mile

6. Question: A five-person family used 12,540 gallons of water during a 30-day period. How many gallons per day did they use?

▼

Answer:
12,540 ÷ 30 = 418 gallons per day

CALCULATIONS

8,650 ÷ 50	19,564 ÷ 292	5,346 ÷ 22	42,842 ÷ 691
173	67	243	62

41,710 ÷ 430	54,488 ÷ 98	67,968 ÷ 708	48,000 ÷ 960
97	556	96	50

352,702 ÷ 59	576,621 ÷ 79	629,760 ÷ 984	8,740,460 ÷ 92
5,978	7,299	640	95,005

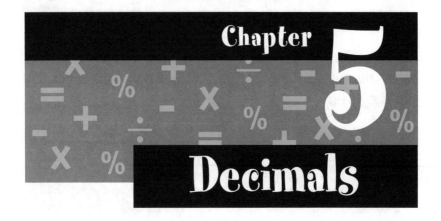

Chapter 5

Decimals

Decimals are a way of expressing numbers less than one. Each decimal place has a different value depending upon its distance from the decimal point. In the end, you have one accurate number. Here are some examples of decimals.

Decimal Point

1	2,	3	4	5.	6	7	8	9
Ten Thousands	Thousands	Hundreds	Tens	Ones	Tenths	Hundredths	Thousandths	Ten Thousandths

Note: .1 = 1/10; .01 = 1/100; .001 = 1/1,000;
.0001 = 1/10,000

EXAMPLE SET 1
ADDING DECIMALS

36.0	1.7	19.6	68.0
+ .9	+ .4	+ 1.6	+ .7
36.9	2.1	21.2	68.7

5.00	5.90	7.38	86.88
+ .37	+ 4.82	+ 2.12	+ .32
5.37	10.72	9.50	87.20

9.000	4.730	3.873	91.262
+ .495	+ 10.203	+ .129	+ 12.726
9.495	14.933	4.002	103.988

1.5000	4.4200	1.5250	8.0275
+ .7463	+ 16.2293	+ .5835	+ .6485
2.2463	20.6493	2.1085	8.6760

EXAMPLE WORD PROBLEM 1

A woman bought an antique gold chain from a jewelry store that weighed 21.75 grams. The following week she bought a gold pendant at an auction that weighed 4.42 grams. She put the pendant on the chain. How much is the total weight?

Solution

21.75
+ 4.42
26.17

The answer is
26.17 grams of gold.

EXAMPLE SET 2
SUBTRACTING DECIMALS

65.0	4.3	84.5	52.0
- .6	- .2	- 5.3	- 12.1
64.4	4.1	79.2	39.9

8.00	9.60	2.55	89.74
- .21	- 2.18	- .93	- 25.41
7.79	7.42	1.62	64.33

5.000	9.780	7.047	97.761
- 4.851	- 5.786	- 5.524	- 50.908
.149	3.994	1.523	46.853

6.3000	5.1400	8.3530	4.9174
- .4357	- .1542	- 6.3423	- 2.6314
5.8643	4.9858	2.0107	2.2860

EXAMPLE WORD PROBLEM 2

A factory produces a piston for an engine. The diameter of the piston is 4.135 inches. After three years of use, the piston was measured again and found to be 4.129 inches. How much wear does the piston show?

Solution

```
  4.135
- 4.129
  .006
```

The answer is .006 inches.

 ## EXAMPLE SET 3
MULTIPLYING DECIMALS

3	47	6.9	91.8
x .5	x .2	x .1	x .4
1.5	9.4	.69	36.72

6	2.2	6.36	72.98
x .56	x .83	x .58	x .93
3.36	1.826	3.6888	67.8714

.9	2.2	.63	7.29
x .56	x .36	x .65	x .86
.504	.792	.4095	6.2694

43	95.6	7	82
x .125	x .204	x .283	x .2846
5.375	19.5024	1.981	23.3372

 ## EXAMPLE WORD PROBLEM 3

Decimals can be used to convert measurements from the English into the metric system. Given that one foot equals .3048 meters, how many meters are there in 40 feet?

Solution

```
    40
x .3048
12.192 meters
```

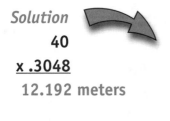

The answer is 12.192 meters.

EXAMPLE SET 4
DIVIDING DECIMALS

.54	14	11.22	1.834
÷ .1	÷ .4	÷ 3.3	÷ .2
5.4	35.0	3.4	9.17

2.52	7.006	3.0022	28.1556
÷ .63	÷ 2.26	÷ .34	÷ .81
4.0	3.1	8.83	34.76

8.736	63.92	6.084	1.287
÷ .48	÷ 7.52	÷ 11.7	÷ .39
18.2	8.5	.52	3.3

4.758	9.375	2.5622	5.0589
÷ .61	÷ .15	÷ 5.57	÷ .73
7.8	62.5	.46	6.93

EXAMPLE WORD PROBLEM 4

A tropical storm passed over a small town. It lasted for 5 whole days and during this time a total of 6.35 inches of rain fell. What was the average number of inches of rain per day during this storm?

Solution
  6.35
   ÷ 5
  1.27 inches

The answer is
1.27 inches of rain.

 ## SPECIAL CASES WITH DECIMALS
MULTIPLYING

When multiplying a decimal number by 10 or 100 or 1000, simply move the decimal point one place to the right for each zero.

7.25 x 10 = 72.5

7.25 x 100 = 725

7.25 x 1000 = 7,250

 ## SPECIAL CASES WITH DECIMALS
DIVIDING

When dividing a decimal number by 10 or 100 or 1000, simply move the decimal point one place to the left for each zero.

7.25 ÷ 10 = .725

7.25 ÷ 100 = .0725

7.25 ÷ 1000 = .00725

READING DECIMAL NUMBERS

1,587,253	one million, five hundred eighty-seven thousand, two hundred fifty-three
120,614	one hundred twenty thousand, six hundred fourteen
49,612	forty-nine thousand, six hundred twelve
5,371	five thousand, three hundred seventy-one
230	two hundred thirty
.8	eight tenths
1.5	one and five tenths
38.6	thirty-eight and six tenths
945.3	nine hundred forty-five and three tenths
2,597.5	two thousand, five hundred ninety-seven and five tenths
.25	twenty-five hundredths
1.62	one and sixty-two hundredths
21.17	twenty-one and seventeen hundredths

389.54   three hundred eighty-nine and fifty-four
          hundredths

8,601.27 eight thousand, six hundred one and
          twenty-seven hundredths

.713   seven hundred thirteen thousandths

2.924   two and nine hundred twenty-four
          thousandths

53.108   fifty-three and one hundred eight
          thousandths

.1175   one thousand one hundred seventy-five
          ten thousandths

5.4833   five and four thousand eight hundred
          thirty-three ten thousandths

76.1845   seventy-six and one thousand eight hundred
          forty-five ten thousandths

Practice Problems*

*Answers to Practice Problems on Pages 72-74

1. A platinum diamond ring has a center stone in it that weighs .875 carats. Several small diamonds that weigh a total of .375 carats surround the center stone. What is the total carat weight of the diamonds in this ring?

2. A wood planer is a machine used to shave the edge off a piece of wood. You start with a piece of wood that is 10 inches thick and pass it through the planer that is set for .125 of an inch. How thick is this piece of wood now?

3. If one inch equals 2.54 centimeters, then how many centimeters does 35 inches equal?

4. There are 28.35 grams in an ounce. How many ounces are there in 567 grams?

5.   A metal supply company sells a wide selection of aluminum sheet metal. Their thinnest sheet is .025 inches. Their thickest sheet is .1875 inches. What is the difference in inches between their thickest and their thinnest aluminum sheet metal?

6.   If one kilogram equals 2.2 pounds, then how many pounds does 24.5 kilograms weigh?

Calculations*

*Answers to Calculations on Page 74

```
  73.0        6.9        1.33       96.14
 + .8        + .3      + 2.29       + .34
```

```
  53.0        2.2        2.52       89.58
 - .7        - .5       - .23      - 87.53
```

```
     2         17        3.16       17.72
 x .4        x .2       x .72       x .63
```

```
  .94         32       2.177     38.6288
 ÷ .1        ÷ .4       ÷ .35       ÷ .56
```

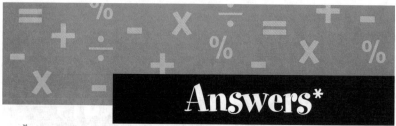

Answers*

*Answers to the preceding Practice Problems and Calculations

1. Question: A platinum diamond ring has a center stone in it that weighs .875 carats. Several small diamonds that together weigh .375 carats surround the center stone. What is the total carat weight of the diamonds in this ring?

Answer:
.875 + .375 = 1.25 carats

2. Question: A wood planer is a machine used to shave the edge off a piece of wood. You start with a piece of wood that is 10 inches thick and pass it through the planer that is set for .125 of an inch. How thick is this piece of wood now?

Answer:
10.0 - .125 = 9.875 inches

3. Question: If one inch equals 2.54 centimeters, then how many centimeters does 35 inches equal?

> *Answer:*
> **35 x 2.54 = 88.9 centimeters**

4. Question: There are 28.35 grams in an ounce. How many ounces are there in 567 grams?

> *Answer:*
> **567 ÷ 28.35 = 20 ounces**

5. Question: A metal supply company sells a wide selection of aluminum sheet metal. Their thinnest sheet is .025 inches. Their thickest sheet is .1875 inches. What is the difference in inches between their thickest and their thinnest aluminum sheet metal?

> *Answer:*
> **.1875 - .025 = .1625 inches**

6. Question: If one kilogram equals 2.2 pounds, then how many pounds does 24.5 kilograms weigh?

> *Answer:*
> **24.5 x 2.2 = 53.9 pounds**

■ CALCULATIONS

73.0	6.9	1.33	96.14
+ .8	+ .3	+ 2.29	+ .34
73.8	7.2	3.62	96.48

53.0	2.2	2.52	89.58
- .7	- .5	- .23	- 87.53
52.3	1.7	2.29	2.05

2	17	3.16	17.72
x .4	x .2	x .72	x .63
.8	3.4	2.2752	11.1636

.94	32	2.177	38.6288
÷ .1	÷ .4	÷ .35	÷ .56
9.4	80.0	6.22	68.98

Chapter 6
Rounding

Rounding means to simplify. In rounding, you drop a few numbers and get an approximation. Rounding involves losing a little information for the sake of convenience. Rounding works like this. First, you choose the decimal place that you want to round to. Then, you look at the digit in the decimal place to the right of it. If that number is 5 or greater then round up. If it is 4 or less, then leave it alone. Lastly, get rid of all the remaining digits to the right of the decimal place that you want. The diagram below names some of the decimal places. Some examples of rounding numbers follow.

Decimal Point

1	2,	3	4	5.	6	7	8	9
Ten Thousands	Thousands	Hundreds	Tens	Ones	Tenths	Hundredths	Thousandths	Ten Thousandths

EXAMPLE SET 1

Round the following number to the nearest tenths.

the place to the right
↓
21.36   =   21.36   ≈   21.4
↑
tenths

The number to the right of the tenths place is a 6. Since 6 is greater than 5, round the 3 up to a 4. Lastly, drop the 6. **Note that the symbol ≈ means approximately equal to.**

Round the following number to the nearest tenths.

the place to the right
↓
21.35   =   21.35   ≈   21.4
↑
tenths

The number to the right of the tenths place is a 5. Since 5 equals 5, round the 3 up to a 4. Lastly, drop the 5. Remember that 5 is the cut-off point.

Round the following number to the nearest tenths.

the place to the right

21.34　　=　　21.3̌4　　≈　　21.3

tenths

The number to the right of the tenths place is a 4. Since 4 is less than 5, leave the 3 alone. Lastly, drop the 4.

EXAMPLE SET 2

Round the following number to the nearest hundreds.

the place to the right

5,721　　=　　5,7̌21　　≈　　5,700

hundreds

The number to the right of the hundreds place is a 2. Since 2 is less than 5, leave the seven alone. Lastly, drop the 2 and the 1 and replace them with zeros.

Round the following number to the nearest tens.

the place to the right
↓

185   =   185   ≈   190

↑
tens

The number to the right of the tens place is a 5. Since 5 equals 5, round the 8 up to a 9. Lastly, drop the 5 and replace it with a zero. Remember that 5 is the cut-off point.

Round the following number to the nearest ones.

the place to the right
↓

3.6   =   3.6   ≈   4

↑
ones

The number to the right of the ones place is a 6. Since 6 is more than 5, round the 3 up to a 4. Lastly, drop the 6.

EXAMPLE SET 3

Round the following number to the nearest tenths.

the place to the right
↓

26.153   =   **26.153**   ≈   26.2
↑
tenths

The number to the right of the tenths place is a 5. Since 5 equals 5, the cut-off point, round the 1 up to a 2. Lastly, drop the 5 and the 3.

Round the following number to the nearest hundredths.

the place to the right
↓

71.3532   =   **71.3532**   ≈   71.35
↑
hundredths

The number to the right of the hundredths place is a 3. Since 3 is less than 5, leave the 5 alone. Lastly, drop the 3 and the 2.

Round the following number to the nearest thousandths.

the place to the right
↓
2.0879   =   2.0879   ≈   2.088
↑
thousandths

The number to the right of the thousandths place is a 9. Since 9 is more than 5, round the 7 up to an 8. Lastly, drop the 9.

EXAMPLE SET 4

The following numbers are rounded to the nearest hundreds.

235 ≈ 200   8,173 ≈ 8,200

47,323 ≈ 47,300   467.62 ≈ 500

1,753 ≈ 1,800   13,578 ≈ 13,600

The following numbers are rounded to the nearest tens.

76 ≈ 80   6,899 ≈ 6,900

875 ≈ 880   961 ≈ 960

65.49 ≈ 70   5,153 ≈ 5,150

The following numbers are rounded to the nearest ones.

23.4 ≈ 23   4.5 ≈ 5

97.28 ≈ 97   12.57 ≈ 13

9.62 ≈ 10   24.367 ≈ 24

The following numbers are rounded to the nearest tenths.

.45 ≈ .5   3.54 ≈ 3.5

.962 ≈ 1   2.11 ≈ 2.1

89.85 ≈ 89.9   5.861 ≈ 5.9

The following numbers are rounded to the nearest hundredths.

.643 ≈ .64   8.987 ≈ 8.99

.1782 ≈ .18   12.765 ≈ 12.77

2.3989 ≈ 2.4   982.915 ≈ 982.92

The following numbers are rounded to the nearest thousandths.

.6121 ≈ .612   3.8175 ≈ 3.818

.1099 ≈ .11   5.8436 ≈ 5.844

79.8484 ≈ 79.848   4.38527 ≈ 4.385

 ROUNDING QUOTIENTS

Whole and decimal numbers can be rounded whether they are presented in ready form or obtained from the results of calculations such as addition, subtraction, multiplication, or division. Rounding is especially useful for division because division can produce what are called repeating numbers. Repeating numbers are when one or a group of numbers repeats forever. The solution is to round that repeating number.

EXAMPLE SET 5

1 ÷ 3 = .333...
rounded off to the nearest tenths is:
$1 \div 3 \approx .3$

2 ÷ 3 = .666...
rounded off to the nearest hundredths is:
$2 \div 3 \approx .67$

1 ÷ 7 = .$\overline{142857}$
rounded of to the nearest thousandths is:
$1 \div 7 \approx .143$

1 ÷ 9 = .111...
rounded off to the nearest hundredths is:
$1 \div 9 \approx .11$

1 ÷ 11 = .0909$\overline{09}$
rounded off to the nearest tenths is:
$1 \div 11 \approx .1$

1 ÷ 12 = .08$\overline{333}$
rounded off to the nearest hundredths is:
$1 \div 12 \approx .08$

1 ÷ 13 = .0$\overline{769230}$
rounded off to the nearest ten thousandths is:
$1 \div 13 \approx .0769$

$1 \div 14 = .\overline{0714285}$
rounded off to the nearest thousands is:
$1 \div 14 \approx .071$

$2 \div 9 = .222...$
rounded off to the nearest hundredths is:
$2 \div 9 \approx .22$

 EXAMPLE SET 6

For each calculation below, the final answers have been rounded off to the nearest hundredths.

76.184 + 53.189 = $129.373 \approx 129.37$

677.6142 - 98.5391 = $579.0751 \approx 579.08$

57.84 x 7.1 = $410.664 \approx 410.66$

353.628 ÷ 85.5 = $4.136 \approx 4.14$

EXAMPLE WORD PROBLEM 1

A formula racecar at the Indianapolis 500 racecourse completed one lap in 38.7572 seconds. Its average rate of speed was 232.215 miles per hour. What are the time and the MPH rounded to the nearest hundredths?

Solution

38.7572 ≈ 38.76 seconds
232.215 ≈ 232.22 MPH

 ## EXAMPLE WORD PROBLEM 2

A farmer from the Midwest sells 2,245 tons of wheat to a broker each month. After 12 months, how many tons of wheat has this farmer sold rounded to the nearest hundreds?

Solution

```
  2,245
 x 12
 26,940
```

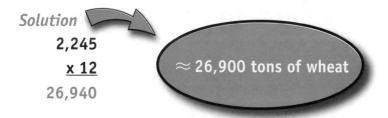

≈ 26,900 tons of wheat

EXAMPLE WORD PROBLEM 3

At We-Sue-Em Law Firm, they recently settled a civil suit for one of their clients. The settlement was in the amount of $44,000. The law firm is entitled to 1/3 of the settlement money for legal services rendered. Rounding to the nearest dollar, how much money will the law firm make?

Solution

$$
\begin{array}{r}
44,000 \\
\div 3 \\
\hline
14,666.666
\end{array}
$$

$\approx \$14,667$

Practice Problems*

*Answers to Practice Problems on Pages 89-90

Round the following numbers to the nearest <u>hundreds</u>.

539	5,780
71,443	796.82
8,554	72,547

Round the following numbers to the nearest <u>tens</u>.

58	5,273
725	814
25.64	3,422

Round the following numbers to the nearest <u>ones</u>.

20.3	6.5
23.34	93.53
5.91	16.318

Round the following numbers to the nearest <u>tenths</u>.

.55	9.27
4.961	5.74
63.25	6.223

Round the following numbers to the nearest <u>hundredths</u>.

.702	1.073
.2316	19.385
5.4389	289.155

Round the following numbers to the nearest <u>thousandths</u>.

.2441	2.6185
.9833	3.7748
27.5832	6.53194

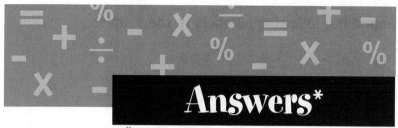

Answers*

*Answers to the preceding Practice Problems

Round the following numbers to the nearest <u>hundreds</u>.

539 ≈ 500   5,780 ≈ 5,800

71,443 ≈ 71,400   796.82 ≈ 800

8,554 ≈ 8,600   72,547 ≈ 72,500

Round the following numbers to the nearest <u>tens</u>.

58 ≈ 60   5,273 ≈ 5,270

725 ≈ 730   814 ≈ 810

25.64 ≈ 30   3,422 ≈ 3,420

Round the following numbers to the nearest <u>ones</u>.

20.3 ≈ 20   6.5 ≈ 7

23.34 ≈ 23   93.53 ≈ 94

5.91 ≈ 6   16.318 ≈ 16

Round the following numbers to the nearest <u>tenths</u>.

.55 ≈ .6   9.27 ≈ 9.3

4.961 ≈ 5.0   5.74 ≈ 5.7

63.25 ≈ 63.3   6.223 ≈ 6.2

Round the following numbers to the nearest <u>hundredths</u>.

.702 ≈ .70   1.073 ≈ 1.07

.2316 ≈ .23   19.385 ≈ 19.39

5.4389 ≈ 5.44   289.155 ≈ 289.16

Round the following numbers to the nearest <u>thousandths</u>.

.2441 ≈ .244   2.6185 ≈ 2.619

.9833 ≈ .983   3.7748 ≈ 3.775

27.5832 ≈ 27.583   6.53194 ≈ 6.532

Chapter 7

Percentages

Percent means per hundred. Percentages are the number of times out of a hundred that something has been observed. Percentages are useful because they allow you to relate a part to a whole. For example, a percentage can be used to express the proportion of voters that support a particular political candidate. The weather person on the television news can say, "There is a 70% chance that it rains tomorrow." Of course there is a 30% chance that it won't, because the chances of it raining plus the chances of it not raining have to equal 100%.

Percentages, decimal numbers, and fractions are all related. A decimal number can be turned into a percentage by multiplying it times 100. For example, the decimal .75 is converted into a percentage by moving the decimal point two places to the right.

.75 x 100 = 75. = 75%

A fraction can be turned into a percentage by dividing the top number, by the bottom number, then multiplying times 100.

$$\frac{3}{4} = 3 \div 4 = .75 = 75\%$$

When working with percentages, there are two basic things that you can do. One is to calculate a percentage. The other is to calculate the percentage of something. The first concern is how to calculate a percentage. Percentages are calculated by dividing the count of the specific group that you are interested in by the count of the whole group.

EXAMPLE WORD PROBLEM 1

A small desert town located in the southwest region receives little rain. Last year it only rained 11 out of the 365 days. In this town, what percent of the days did it rain last year? Round to the nearest percent.

Solution

$$\begin{array}{r} 11 \\ \div\ 365 \\ \hline \approx .03 \end{array}$$

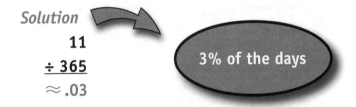

3% of the days

EXAMPLE WORD PROBLEM 2

Moe Power is a mayoral candidate who wants to know how well his election campaign is going. So, he conducts a survey. Out of 2,000 eligible voters surveyed, 1,260 said that they planned to vote for Moe in the upcoming election. What percent of the vote is this?

Solution

1,260
÷ 2000
.63

63% of the vote

 EXAMPLE WORD PROBLEM 3

The enrollment statistics at State College shows that out of the 10,400 students enrolled, 5,460 are women. What is the percentage of women? Also, What is the percentage of men? Round to the nearest tenth of a percent.

Solution

5,460
÷ 10,400
.525

52.5% are women

47.5% are men

100
- 52.5
47.5

■ EXAMPLE WORD PROBLEM 4

On the Blue Wave cruise ship, passengers have their choice between a standard stateroom and a deluxe stateroom. There are 242 standard staterooms and 96 deluxe. What percent of the staterooms are deluxe? Round to the nearest percent.

Solution

242
+ 96
338 total staterooms

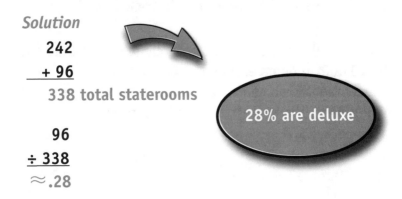

28% are deluxe

96
÷ 338
≈ .28

EXAMPLE WORD PROBLEM 5

In a recent survey of 8,000 coffee drinkers, 960 of them preferred decaffeinated coffee to regular coffee. What percent of coffee drinkers prefer decaffeinated?

Solution

```
    960
÷ 8,000
    .12
```

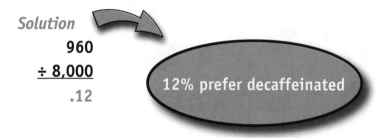

12% prefer decaffeinated

In order to calculate the percentage of something, you first have to convert the percentage into decimal form. A percentage is converted into decimal form by dividing it by 100. A simple way to divide by 100 is to move the decimal point two places to the left.

75% = .75

After converting the percentage into decimal form, the second step is to multiply your quantity of interest times your decimal number. Here are some examples of calculating the percentage of something.

EXAMPLE WORD PROBLEM 6

A businesswoman has a meal at a nice restaurant. She likes to tip 20% when the service is excellent. How much money will she tip on a $35 meal?

Solution

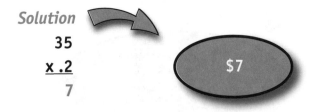

```
    35
  x .2
    7
```

$7

EXAMPLE WORD PROBLEM 7

A money market account at First Bank pays 1.93% annual interest. If you put $10,000 in this account for one year, how much interest will you earn?

Solution

1.93% = .0193

$193 interest

```
  10,000
  x .0193
     193
```

EXAMPLE WORD PROBLEM 8

A small tropical country imports 35% of its bananas. This country consumes a total of 685,000 pounds of bananas each year. How many pounds of bananas does this county import each year?

Solution

685,000
x .35
239,750

239,750 pounds

 EXAMPLE WORD PROBLEM 9

A horse rancher has land located in the Great Plains region, where she owns 1,200 horses. She has different breeds of horses but her favorite is the mustang. 65% of her horses are mustangs. How many mustangs does the rancher own?

Solution

1,200
x .65
780

780 mustangs

EXAMPLE WORD PROBLEM 10

In one suburban city, 42% of the people are homeowners. This city has a population of 90,000 people. How many homeowners are there in this city?

Solution

90,000
x .42
37,800

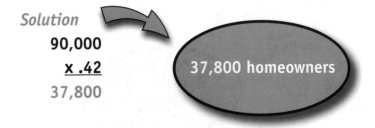

37,800 homeowners

*Answers to Practice Problems on Pages 102-106

1. Nevada is a fast growing state. In 1990 the population was 770,280 people. In 2000 the population was 1,394,440. What is the percentage of population growth from 1990 to 2000? Round to the nearest percent.

2. A cell phone company is having a sale. A phone that normally sells for $160 is on sale for $120. What is the percent saving?

3. Nine games into the season, a professional basketball player has taken 105 shots and made 51 of them. What is this player's shooting percentage? Round to the nearest tenth of a percent.

4. A television commercial for chewing gum claims that 4 out of 5 dentists recommend their gum to their patients who chew gum. What percent of dentists recommend this gum?

5. Last week, a professional football team had 50,000 fans attend their game. This week, 65,000 fans attended their game. What is the percent increase in attendance from last week to this week? Round to the nearest percent.

6. A venture capitalist invested $500,000 into a business endeavor that paid back all of the investment plus a profit of $80,000. What is the percent profit from this deal?

7. Charge-U credit card company charges 12.9% annual interest. If you maintain a balance of $3,000 for one year, what would be your interest charge?

8. A traveler likes to tip 15% on taxi car rides. If the taxi fair amounts to $28 how much money should the traveler tip?

9. A math competency exam was administered to 1,375 students. 84% of the students passed the math exam. How many students passed the math exam.

10. In a recent survey, 57% of the respondents reported owning a dog. If there were 300 people in this survey, how many were dog owners?

11. Old Jack Whiskey is 40% alcohol by volume. How many quarts of alcohol are there in 12 quarts of this whiskey?

12. Wholesome Elementary School has an enrollment of 950 students. From testing, it has been determined that 4% of the students have some type of learning disability. How many students in this school have a learning disability?

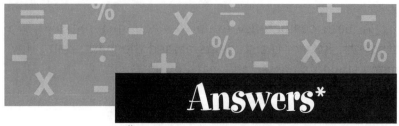

Answers*

*Answers to the preceding Practice Problems

1. Question: Nevada is a fast growing state. In 1990 the population was 770,280 people. In 2000 the population was 1,394,440. What is the percentage of population growth from 1990 to 2000? Round to the nearest percent.

Answer:
1,394,440 - 770,280 = 624,160

624,160 ÷ 1,394,440 ≈ .45 = 45% growth

2. Question: A cell phone company is having a sale. A phone that normally sells for $160 is on sale for $120. What is the percent saving?

Answer:
160 − 120 = 40
40 ÷ 160 = .25 = 25% saving

3. Question: Nine games into the season, a professional basketball player has taken 105 shots and made 51 of them. What is this player's shooting percentage? Round to the nearest tenth of a percent.

Answer:
51 ÷ 105 = .486 = 48.6% shooting

4. Question: A television commercial for chewing gum claims that 4 out of 5 dentists recommend their gum to their patients who chew gum. What percent of dentists recommend this gum?

Answer:
4 ÷ 5 = .8 = 80% recommend

5. Question: Last week, a professional football team had 50,000 fans attend their game. This week, 65,000 fans attended their game. What is the percent increase in attendance from last week to this week? Round to the nearest percent.

> *Answer:*
> 65,000 - 50,000 = 15,000
> 15,000 ÷ 65,000 ≈ .23 = 23% increase

6. Question: A venture capitalist invested $500,000 into a business endeavor that paid back all of the investment plus a profit of $80,000. What is the percent profit from this deal?

> *Answer:*
> 80,000 ÷ 500,000 = .16 = 16% profit

7. Question: Charge-U credit card company charges 12.9% annual interest. If you maintain a balance of $3,000 for one year, what would be your interest charge?

> *Answer:*
> 12.9% in decimal form is .129
> 3,000 x .129 = $387 interest

8. Question: A traveler likes to tip 15% on taxi car rides. If the taxi fare amounts to $28 how much money should the traveler tip?

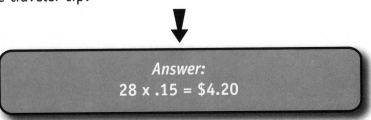

Answer:
28 x .15 = $4.20

9. Question: A math competency exam was administered to 1,375 students. 84% of the students passed the math exam. How many students passed the math exam?

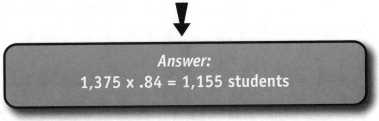

Answer:
1,375 x .84 = 1,155 students

10. Question: In a recent survey, 57% of the respondents reported owning a dog. If there were 300 people in this survey, how many were dog owners?

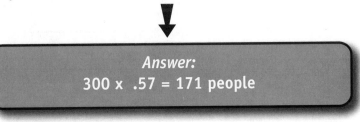

Answer:
300 x .57 = 171 people

11. Question: Old Jack Whiskey is 40% alcohol by volume. How many quarts of alcohol are there in 12 quarts of this whiskey?

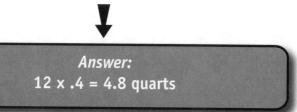

Answer:
12 x .4 = 4.8 quarts

12. Question: Wholesome Elementary School has an enrollment of 950 students. From testing, it has been determined that 4% of the students have some type of learning disability. How many students in this school have a learning disability?

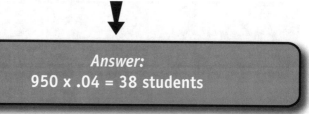

Answer:
950 x .04 = 38 students

Chapter 8

Fractions

A fraction is a part of something. A fraction allows you to express things in less than whole units. The top part of a fraction is called the numerator and the bottom part of a fraction is called the denominator. A fraction can be expressed in terms of a decimal or a percentage.

A fraction can be converted into a decimal number. In order to convert a fraction into a decimal divide the top number by the bottom number. For example:

$$\frac{3}{4} = 3 \div 4 = .75$$

A fraction can be converted into a percentage by first changing it into a decimal number, then multiplying it times 100. For example:

$$\frac{2}{5} = 2 \div 5 = .4 = 40\%$$

ADDITION OF FRACTIONS

When adding or subtracting fractions you need to find the lowest common denominator. The lowest common denominator is the smallest number that can be divided by both denominators. First, find the lowest common denominator between the two fractions. In the case below, the common denominator is 12 because both 3 and 4 divide evenly into 12. In order to change the two fractions into fractions that have a common denominator, multiply the top and bottom of the first fraction by 4. Then, multiply the top and bottom of the second fraction by 3. Lastly, add the two fractions together. 4 twelfths plus 3 twelfths equals 7 twelfths.

$$\frac{1}{3} + \frac{1}{4} = \frac{4 \times 1}{4 \times 3} + \frac{3 \times 1}{3 \times 4} = \frac{4}{12} + \frac{3}{12} = \frac{7}{12}$$

Instead of finding a common denominator, an alternative is to convert each fraction into a decimal by dividing, then performing the addition or subtraction (as the case warrants). *In this chapter, all calculations are rounded off to the nearest thousandths.*

EXAMPLE SET 1

$\dfrac{1}{3}+\dfrac{1}{4}=?$

$\dfrac{1}{3}=.333$ $\dfrac{1}{4}=.25$

➤

$$\begin{array}{r} .333 \\ +\,.250 \\ \hline .583 \end{array}$$

$\dfrac{1}{2}+\dfrac{3}{8}=?$

$\dfrac{1}{2}=.5$ $\dfrac{3}{8}=.375$

➤

$$\begin{array}{r} .500 \\ +\,.375 \\ \hline .875 \end{array}$$

$\dfrac{3}{7}+\dfrac{2}{3}=?$

$\dfrac{3}{7}=.429$ $\dfrac{2}{3}=.667$

➤

$$\begin{array}{r} .429 \\ +\,.667 \\ \hline 1.096 \end{array}$$

$3\dfrac{2}{9}+1\dfrac{5}{12}=?$

$3\dfrac{2}{9}=3.222$ $1\dfrac{5}{12}=1.417$

➤

$$\begin{array}{r} 3.222 \\ +\,1.417 \\ \hline 4.639 \end{array}$$

 EXAMPLE WORD PROBLEM 1

The Rusty Hook Seafood Store has fresh crabs for sale. A customer picks out one crab that weighs 2 and 5/8 pounds and another one that weighs 2 and 1/4 pounds. How much do the two crabs weigh together?

Solution

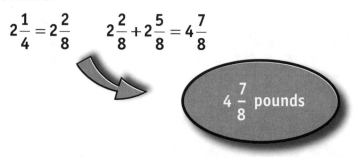

$$2\frac{1}{4} = 2\frac{2}{8} \qquad 2\frac{2}{8} + 2\frac{5}{8} = 4\frac{7}{8}$$

$$4\frac{7}{8} \text{ pounds}$$

■ SUBTRACTION OF FRACTIONS

Just like adding fractions, when subtracting fractions, you need to find a common denominator. For example:

$$\frac{3}{5} - \frac{2}{7} = \frac{7 \times 3}{7 \times 5} - \frac{5 \times 2}{5 \times 7} = \frac{21}{35} - \frac{10}{35} = \frac{11}{35}$$

Instead of finding a common denominator, an alternative is to convert each fraction into a decimal by dividing, then performing the subtraction.

EXAMPLE SET 2

$$\frac{3}{5} - \frac{2}{7} = ?$$

$$\frac{3}{5} = .6 \qquad \frac{2}{7} = .286$$

$$
\begin{array}{r}
.600 \\
-\ .286 \\
\hline
.314
\end{array}
$$

$$\frac{5}{8} - \frac{1}{3} = ?$$

$$\frac{5}{8} = .625 \qquad \frac{1}{3} = .333$$

$$
\begin{array}{r}
.625 \\
-\ .333 \\
\hline
.292
\end{array}
$$

$$\frac{4}{5} - \frac{8}{11} = ?$$

$$\frac{4}{5} = .8 \qquad \frac{8}{11} = .727$$

$$
\begin{array}{r}
.800 \\
-\ .727 \\
\hline
.073
\end{array}
$$

$$\frac{2}{10} - \frac{1}{6} = ?$$

$$\frac{2}{10} = .2 \qquad \frac{1}{6} = .167$$

$$
\begin{array}{r}
.200 \\
-\ .167 \\
\hline
.033
\end{array}
$$

$$5\frac{2}{3} - 2\frac{3}{4} = ?$$

$$5\frac{2}{3} = 5.667 \qquad 2\frac{3}{4} = 2.75$$

$$
\begin{array}{r}
5.667 \\
-\ 2.750 \\
\hline
2.917
\end{array}
$$

EXAMPLE WORD PROBLEM 2

What is the difference between 1/4 of a gallon and 1/5 of a gallon?

Solution

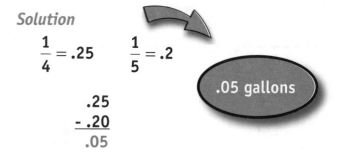

$$\frac{1}{4} = .25 \qquad \frac{1}{5} = .2$$

.05 gallons

$$
\begin{array}{r}
.25 \\
- .20 \\
\hline
.05
\end{array}
$$

 MULTIPLICATION OF FRACTIONS

When multiplying fractions, multiply the top times the top, and the bottom times the bottom. For example:

$$\frac{3}{4} \times \frac{5}{7} = \frac{3 \times 5}{4 \times 7} = \frac{15}{28}$$

A mixed number is a whole number and a fraction. With mixed numbers, first convert the mixed number into a fraction. You do this by multiplying the whole number times the bottom of the fraction and adding this to the number at the top of the fraction. Lastly, multiply the top times the top and the bottom times the bottom. For example:

$$4\frac{1}{2} \times 1\frac{7}{8} = \frac{9}{2} \times \frac{15}{8} = \frac{135}{16}$$

The answer can be left in this fraction form, or it can be converted into a decimal number by dividing the top number by the bottom number.

 EXAMPLE SET 3

$$\frac{1}{4} \times \frac{13}{16} = \frac{13}{64} = .203$$

$$\frac{2}{9} \times \frac{3}{8} = \frac{6}{72} = \frac{1}{12} = .083$$

$$6\frac{1}{2} \times 4\frac{7}{20} = \frac{13}{2} \times \frac{87}{20} = \frac{1,131}{40} = 28.275$$

 EXAMPLE WORD PROBLEM 3

A casserole recipe calls for 1 and 1/2 tablespoons of crushed garlic. If a cook wants to make 2 and 1/2 times the recipe, how many tablespoons of garlic does the cook need?

Solution

$$1\frac{1}{2} \times 2\frac{1}{2} = \frac{3}{2} \times \frac{5}{2} = \frac{15}{4} = 3\frac{3}{4}$$

$3\frac{3}{4}$ tablespoons of garlic

DIVISION OF FRACTIONS

When dividing fractions, multiply the first fraction times the second fraction turned upside down. The second fraction turned upside down is called the reciprocal. For example:

$$\frac{3}{4} \div \frac{2}{5} = \frac{3}{4} \times \frac{5}{2} = \frac{15}{8} = 1.875$$

With mixed numbers: First convert the mixed number into a fraction. Then, divide by multiplying the first fraction times the second fraction turned upside down.

$$8\frac{1}{4} \div 2\frac{2}{3} = \frac{33}{4} \div \frac{8}{3} = \frac{33}{4} \times \frac{3}{8} = \frac{99}{32} = 3.094$$

EXAMPLE SET 4

$$\frac{9}{16} \div \frac{1}{4} = \frac{9}{16} \times \frac{4}{1} = \frac{36}{16} = 2.25$$

$$\frac{15}{21} \div \frac{2}{9} = \frac{15}{21} \times \frac{9}{2} = \frac{135}{42} = 3.214$$

$$12\frac{3}{4} \div 3\frac{1}{6} = \frac{51}{4} \div \frac{19}{6} = \frac{51}{4} \times \frac{6}{19} = \frac{306}{76} = 4.026$$

 EXAMPLE WORD PROBLEM 4

A chef at a restaurant is making fruit cocktails. It takes 2/5 of a pound of fruit for each cocktail. How many servings of fruit cocktail can the chef make with 35 pounds of fruit?

Solution

$$35 \div \frac{2}{5} = \frac{35}{1} \div \frac{2}{5} = \frac{35}{1} \times \frac{5}{2} = \frac{175}{2} = 87.5$$

87.5 servings of fruit cocktail

Practice Problems*

*Answers to Practice Problems on Pages 118-119

$$\frac{2}{3} + \frac{3}{16} =$$

$$\frac{1}{2} + \frac{7}{8} =$$

$$4\frac{8}{9} + 3\frac{1}{7} =$$

$$\frac{5}{8} - \frac{7}{13} =$$

$$\frac{5}{12} - \frac{1}{5} =$$

$$7\frac{1}{4} - 3\frac{3}{8} =$$

$$\frac{1}{5} \times \frac{7}{11} =$$

$$\frac{4}{9} \times \frac{3}{4} =$$

$$6\frac{1}{3} \times 3\frac{1}{4} =$$

$$\frac{5}{8} \div \frac{2}{3} =$$

$$\frac{13}{16} \div \frac{4}{7} =$$

$$6\frac{1}{2} \div 2\frac{1}{4} =$$

*Answers to the preceding Practice Problems

$\frac{2}{3} + \frac{3}{16} =$

$\frac{2}{3} = .667$ $\frac{3}{16} = .188$ ▶

```
  .667
+ .188
  .855
```

$\frac{1}{2} + \frac{7}{8} =$

$\frac{1}{2} = .5$ $\frac{7}{8} = .875$ ▶

```
  .500
+ .875
 1.375
```

$4\frac{8}{9} + 3\frac{1}{7} =$

$4\frac{8}{9} = 4.889$ $3\frac{1}{7} = 3.143$ ▶

```
  4.889
+ 3.143
  8.032
```

$\frac{5}{8} - \frac{7}{13} =$

$\frac{5}{8} = .625$ $\frac{7}{13} = .538$ ▶

```
  .625
- .538
  .087
```

$$\frac{5}{12} - \frac{1}{5} =$$

$$\frac{5}{12} = .417 \qquad \frac{1}{5} = .2 \qquad \blacktriangleright$$

$$\begin{array}{r} .417 \\ - .200 \\ \hline .217 \end{array}$$

$$7\frac{1}{4} - 3\frac{3}{8} =$$

$$7\frac{1}{4} = 7.25 \qquad 3\frac{3}{8} = 3.375 \qquad \blacktriangleright$$

$$\begin{array}{r} 7.250 \\ - 3.375 \\ \hline 3.875 \end{array}$$

$$\frac{1}{5} \times \frac{7}{11} = \frac{7}{55} = .127$$

$$\frac{4}{9} \times \frac{3}{4} = \frac{12}{36} = \frac{1}{3} = .333$$

$$6\frac{1}{3} \times 3\frac{1}{4} = \frac{19}{3} \times \frac{13}{4} = \frac{247}{12} = 20.583$$

$$\frac{5}{8} \div \frac{2}{3} = \frac{5}{8} \times \frac{3}{2} = \frac{15}{16} = .938$$

$$\frac{13}{16} \div \frac{4}{7} = \frac{13}{16} \times \frac{7}{4} = \frac{91}{64} = 1.422$$

$$6\frac{1}{2} \div 2\frac{1}{4} = \frac{13}{2} \div \frac{9}{4} = \frac{13}{2} \times \frac{4}{9} = \frac{52}{18} = 2.889$$

Chapter 9

Averages

Averages tell us what is usual, normal, and to be expected. They describe. They are measurements of central tendency. Averages take into consideration all of the numbers in a data set. There are three different types of averages. They are the mean, median, and mode.

The mean is the average score. The mean is what we typically think of as the average. In order to calculate the mean of a set of scores, add up all of the scores and then divide by the number of scores.

The median is the middle score in a group. First, rank order the scores from smallest to largest. If you have an odd number of scores, then the middle score is the median. If you have an even number of scores, then add the two middle scores together and divide by 2. The median score is at the 50th percentile, in other words, it is higher than 50% of all the scores.

The mode is the most frequently occurring score.

EXAMPLE 1

The number of runs scored by a baseball team was recorded for an 8 game period.

Scores: { 3, 2, 5, 6, 2, 2, 1, 3 }

Mean: Add up the scores, then divide by the total number of scores. The sum of scores equals 24 and there are 8 scores.

$$\text{mean} = \frac{24}{8} = 3$$

Median: First, rank order the scores from smallest to largest. Since there is an even number of scores, the median is between the two middle scores.

Scores in Order: { 1, 2, 2, 2, 3, 3, 5, 6 }

$$\text{median} = \frac{2+3}{2} = \frac{5}{2} = 2.5 \qquad \textbf{median}$$
2.5

Mode: The mode is 2 because it occurs more often than any other score.

EXAMPLE 2

In a survey on foot size, 8 elementary high school boys had their feet measured.

Scores: { 7, 6, 6, 10, 5, 7, 6, 9 }

Mean: Add up the scores, then divide by the total number of scores. The sum of scores equals 56 and there are 8 scores.

$$\text{mean} = \frac{56}{8} = 7$$

Median: First, rank order the scores from smallest to largest. Since there is an even number of scores, the median is between the two middle scores.

Scores in Order: { 5, 6, 6, 6, 7, 7, 9, 10 }

$$\text{median} = \frac{6 + 7}{2} = \frac{13}{2} = 6.5 \qquad \text{median} \\ 6.5$$

Mode: The mode is 6 because it occurs more often than any other score.

EXAMPLE 3

Employees at a manufacturing company took a job satisfaction survey. Low scores indicate low job satisfaction and high scores indicate high job satisfaction.

Scores: { 3, 4, 6, 7, 2, 7, 4, 8, 5, 9, 4, 1 }

Mean: Add up the scores, then divide by the total number of scores. The sum of scores equals 60 and there are 12 scores.

$$\text{mean} = \frac{60}{12} = 5$$

Median: First, rank order the scores from smallest to largest. Since there is an even number of scores, the median is between the two middle scores.

Scores in Order: { 1, 2, 3, 4, 4, 4, 5, 6, 7, 7, 8, 9 }

$$\text{median} = \frac{4 + 5}{2} = \frac{9}{2} = 4.5$$

median
4.5

Mode: The mode is 4 because it occurs more often than any other score.

EXAMPLE 4

A basketball team had a contest to see how many baskets each team member could make during a two-minute period.

Scores: { 7, 10, 9, 5, 7, 6, 8, 7, 9, 10, 8 }

Mean: Add up the scores, then divide by the total number of scores. The sum of scores equals 86 and there are 11 scores.

$$\text{mean} = \frac{86}{11} = 7.82$$

Median: First, rank order the scores from smallest to largest. Since there is an odd number of scores, the median is the score in the middle.

Scores in Order: { 5, 6, 7, 7, 7, 8, 8, 9, 9, 10, 10 }

median = 8   median
                                8

Mode: The mode is 7 because it occurs more often than any other score.

EXAMPLE WORD PROBLEM 1

The yearly rainfall totals for a particular county in the Midwest was recorded for 5 consecutive years. The yearly rainfall amounts are listed below. What was the average yearly rainfall?

Solution

year 1	19
year 2	28
year 3	22
year 4	35
year 5	+ 23
	127

$$\frac{127}{5} = 25.4 \text{ inches of rain}$$

EXAMPLE WORD PROBLEM 2

A crate full of 50 large apples weighs 26.5 pounds, which converts to 424 ounces. What is the average weight of an apple in this crate (in ounces)?

Solution

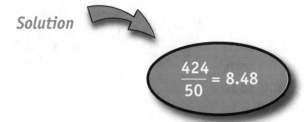

$$\frac{424}{50} = 8.48$$

EXAMPLE WORD PROBLEM 3

Last year, one particular family spent a total of $871.80 on telephone bills. What is the average monthly bill?

Solution

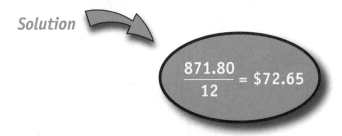

$$\frac{871.80}{12} = \$72.65$$

 EXAMPLE WORD PROBLEM 4

One aluminum-recycling center processes tons of aluminum each month. The amount of aluminum processed for a six-month period is displayed below. What is the average amount recycled per month?

Solution

month 1	18,000 tons
month 2	28,000 tons
month 3	41,000 tons
month 4	24,000 tons
month 5	34,000 tons
month 6	+ 23,000 tons
	168,000

$$\frac{168,000}{6} = 28,000 \text{ tons}$$

EXAMPLE WORD PROBLEM 5

At a recent college track and field event, 8 women athletes ran the 100-meter dash. The total time for all 8 athletes was 97.12 seconds. What was the average time for an athlete in this race?

Solution

$$\frac{97.12}{8} = 12.14 \text{ seconds}$$

 EXAMPLE WORD PROBLEM 6

The annual income of one particular independent contractor varies depending upon how good business was that year. Using the earnings for the last 4 years listed below, calculate his average income per year.

Solution

year 1	**$51,000**
year 2	**$47,000**
year 3	**$53,000**
year 4	**+ $42,000**
	$193,000

$$\frac{193,000}{4} = \$48,250$$

Practice Problems*

*Answers to Practice Problems on Pages 131-136

1. Ten students in an elementary school completed an extra credit geography project. The maximum number of points possible is 10. The scores are: 8, 5, 7, 5, 10, 6, 8, 9, 8, and 4. What are the mean, median, and mode?

2. At a particular college, the school record for the men's one-mile relay race is 3 minutes and 19 seconds. Given that the race is 4 laps long, what is the average time per lap? *(Hint: Convert minutes into seconds.)*

3. A county in one Midwestern state received a total of 225 inches of snow during the past 10 years. What is the average yearly snowfall?

4. Five men were measured for height. The measurements are: 5-feet, 6 inches; 5-feet, 10 inches; 5-feet, 11 inches; 5-feet, 8 inches; 5-feet, 5 inches. What is the average height of this group? *(Hint: Convert feet to inches.)*

5. A warm spell has settled on a small western town. The high temperatures there last week were: 81, 83, 85, 87, 88, 88, and 90 degrees. What was the average high temperature?

6. A student gets his report card in the mail and wants to figure out his grade-point-average (or GPA). He knows that an A = 4 points, B = 3 points, C = 2 points, and D = 1 point. This semester, the student earned two As, two Bs, and a C. What is this student's GPA for this semester?

7. Over the last 20 college basketball games, a woman athlete has scored a total of 170 points. What is her average number of points per game?

8. Attendance at The First Non-Denominational Church fluctuates from one week to the next. For the last 4 weeks, the attendance numbers are: 440, 418, 397, and 485. What is the average attendance per week?

9. Twenty-five high school students took a test in English class. The maximum number of points possible is 100. The sum of the scores for the 25 students was 2,065. What is the average test score?

10. A baseball player gets 43 hits in 150 at bats. Rounding to the nearest thousandths, what is the player's batting average?

Answers*

1. Question: Ten students in an elementary school completed an extra credit geography project. The maximum number of points possible is 10. The scores are: 8, 5, 7, 5, 10, 6, 8, 9, 8, and 4. What are the mean, median, and mode?

Answers: 8 + 5 + 7 + 5 + 10 + 6 + 8 + 9 + 8 + 4 = 70

$$\text{mean} = \frac{70}{10} = 7$$

Median: First, rank order the scores from smallest to largest. Since there is an even number of scores, the median is between the two middle scores.

Scores in Order: { 4, 5, 5, 6, 7, 8, 8, 8, 9, 10}

$$\text{median} = \frac{7+8}{2} = \frac{15}{2} = 7.5$$ **median**
7.5

Mode: The mode is 8 because it occurs more often than any other score.

2. Question: At a particular college, the school record for the men's one-mile relay race is 3 minutes and 19 seconds. Given that the race is 4 laps long, what is the average time per lap? *(Hint: Convert minutes into seconds.)*

Answer:
Since there are 60 seconds in one minute, there are 180 seconds in 3 minutes.
180 seconds plus 19 seconds equals 199 seconds.

$$\frac{199}{4} = 49.75 \text{ seconds}$$

3. Question: A county in one Midwestern state received a total of 225 inches of snow during the past 10 years. What is the average yearly snowfall?

Answer:
$$\frac{225}{10} = 22.5 \text{ inches}$$

4. Question: Five men were measured for height. The measurements are: 5-feet, 6 inches; 5-feet, 10 inches; 5-feet, 11 inches; 5-feet, 8 inches; 5-feet, 5 inches. What is the average height of this group? *(Hint: Convert feet to inches.)*

Answer:
First, convert each height into inches.
Since there are 60 inches in 5 feet, the heights in inches are 66, 70, 71, 68, and 65.
The sum is 340 inches.

$$\frac{340}{5} = 68 \text{ inches, which equals}$$
5 feet, 8 inches

5. Question: A warm spell has settled on a small western town. The high temperatures there last week were: 81, 83, 85, 87, 88, 88, and 90 degrees. What was the average high temperature?

Answer:
The sum of measurements is 602.

$$\frac{602}{7} = 86 \text{ degrees}$$

6. Question: A student gets his report card in the mail and wants to figure out his grade-point-average (or GPA). He knows that an A = 4 points, B = 3 points, C = 2 points, and D = 1 point. This semester, the student earned two As, two Bs, and a C. What is this student's GPA for this semester?

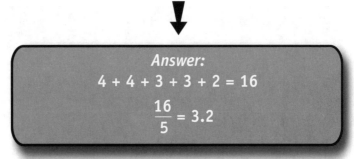

Answer:
4 + 4 + 3 + 3 + 2 = 16

$$\frac{16}{5} = 3.2$$

7. Question: Over the last 20 college basketball games, a woman athlete has scored a total of 170 points. What is her average number of points per game?

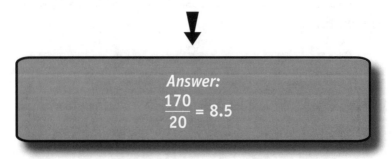

Answer:
$$\frac{170}{20} = 8.5$$

8. Question: Attendance at The First Non-Denominational Church fluctuates from one week to the next. For the last 4 weeks, the attendance numbers are: 440, 418, 397, and 485. What is the average attendance per week?

Answer:

$$440 + 418 + 397 + 485 = 1,740$$

$$\frac{1,740}{4} = 435 \text{ people}$$

9. Question: Twenty-five high school students took a test in English class. The maximum number of points possible is 100. The sum of the scores for the 25 students was 2,065. What is the average test score?

Answer:

$$\frac{2,065}{25} = 82.6 \text{ points}$$

10. Question: A baseball player gets 43 hits in 150 at bats. Rounding to the nearest thousandths, what is the player's batting average?

Answer:

$$\frac{43}{150} = .287$$

Chapter 10

Sales Tax

Sales tax is a fact of life that you have to pay at the checkout stand. Sales tax is levied on the sales of certain goods for the support of the government. It is usually expressed in terms of a percentage. If you are a seller, you need to figure out how much to charge. If you are a buyer, you need to figure out how much it will cost you. The first step in calculating sales tax is to convert the tax into decimal form. The second step is to multiply the price times the tax.

A percentage is turned into a decimal number by dividing it by 100. An easy way to divide any number by 100 is to simply move the decimal point two places to the left. Here are some examples of converting a percentage into decimal form followed by some examples of calculating sales tax.

8% = .08   5% = .05

4.5% = .045   6.25% = .0625

EXAMPLE WORD PROBLEM 1

A customer is considering buying a new car from Slick-Guys Motors. The price of the model that the customer wants is $25,300. What is the total price if the sales tax is 6%?

Solution

6% in decimal form is .06.

Multiply $25,300 by .06 to find the sales tax.

> **25,300 price**
> **x .06**
> $1,518

➤ The sales tax is $1,518.

Add the sales tax to the price.

> **25,300 price**
> **+ 1,518 tax**
> $26,818

The total price is
$26,818.

EXAMPLE WORD PROBLEM 2

A person shopping at the mall finds just the right pair of athletic shoes. The price of the shoes is $42. What is the total price if the sales tax is 4%?

Solution

4% in decimal form is .04.

Multiply $42 by .04 to find the sales tax.

     42 price
   x .04
    $1.68   ►►   The sales tax is $1.68.

Add the sales tax to the price.

   42.00 price
  + 1.68 tax
   $43.68

The total price is
$43.68.

EXAMPLE WORD PROBLEM 3

In preparation for a child's birthday party, a parent buys a dozen balloons. The balloons cost $6. What is the total price if the sales tax is 5.3%?

Solution

5.3% in decimal form is .053.

Multiply $6 by .053 to find the sales tax.

```
      6 price
   x .053
    .318 ≈ 32
```
 The sales tax is 32 cents.

Add the sales tax to the price.

```
   6.00 price
 + .32 tax
  $6.32
```

The total price is $6.32.

EXAMPLE WORD PROBLEM 4

The bill for a good meal at Fine Food Restaurant comes to $17.50. What is the total price if the sales tax is 7%?

Solution

7% in decimal form is .07.

Multiply $17.50 by .07 to find the sales tax.

    17.50 price
    x .07
    $1.225 ≈ $1.23 ➡️ The sales tax is $1.23.

Add the sales tax to the price.

    17.50 price
    + 1.23 tax
    $18.73

The total price is $18.73.

EXAMPLE WORD PROBLEM 5

After careful research, a person has decided to buy a Samsox computer. The computer sells for $1,200. What is the total price if the sale tax is 8.25%?

Solution

8.25% in decimal form is .0825.

Multiply $1,200 by .0825 to find the sales tax.

1,200 price
x .0825
   $99      The sales tax is $99.

Add the sales tax to the price.

1,200 price
 + 99 tax
$1,299

The total price is $1,299.

EXAMPLE WORD PROBLEM 6

While shopping in a sporting goods store, a man found a hat that fit him perfectly. The hat sells for $34. What is the total price if the sales tax is 6.5%?

Solution

6.5% in decimal form is .065.

Multiply $34 by .065 to find the sales tax.

     34 price
x .065
   $2.21   ▶   The sales tax is $2.21.

Add the sales tax to the price.

  34.00 price
 + 2.21 tax
  $36.21

The total price is $36.21.

EXAMPLE WORD PROBLEM 7

A wood fence will cost a homeowner $1,850 plus tax. What is the total price if the sales tax is 5%?

Solution

5% in decimal form is .05.

Multiply $1,850 by .05 to find the sales tax.

> 1,850 price
> x .05
> $92.50

    The sales tax is $92.50.

Add the sales tax to the price.

> 1,850.00 price
> + 92.50 tax
> $1,942.50

The total price is $1,942.50.

EXAMPLE WORD PROBLEM 8

At Music-Man Music Store they have every type of music CD that you can imagine. The CDs sell for $20 a piece. What is the total price for one if the sales tax is 5.75%?

Solution

5.75% in decimal form is .0575.

Multiply $20 by .0575 to find the sales tax.

```
        20 price
      x .0575
        $1.15
```
➡ The sales tax is $1.15.

Add the sales tax to the price.

```
      20.00 price
    +  1.15 tax
      $21.15
```

The total price is $21.15.

EXAMPLE WORD PROBLEM 9

A photographer was happy to find an exceptional quality camera at the price of $320. What is the total price if the sales tax is 7%?

Solution

7% in decimal form is .07.

Multiply $320 by .07 to find the sales tax.

> **320** price
> **x .07**
> $22.40 ▶ The sales tax is $22.40.

Add the sales tax to the price.

> **320.00** price
> **+ 22.40** tax
> $342.40

The total price is $342.40.

EXAMPLE WORD PROBLEM 10

A secretary makes a trip to a store called Everything-Office to buy some office supplies. Everything in the shopping basket totals $91.25. What is the total price if the sales tax is 4%?

Solution

4% in decimal form is .04.

Multiply $91.25 by .04 to find the sales tax.

    91.25 price
     x .04
    $3.65      The sales tax is $3.65.

Add the sales tax to the price.

    91.25 price
    + 3.65 tax
    $94.90

The total price is $94.90.

Practice Problems*

*Answers to Practice Problems on Pages 150-154

1. Golf is a serious game and so is the cost of the equipment. One popular set of golf clubs sells for $250. What is the total price if the sales tax is 4%?

2. At Bright-Lights Electronics, they have a new television for $395. What is the total price if the sales tax is 7%?

3. One night at a comfortable hotel costs a traveler $85.60 plus tax. What is the total price if the tax is 5.6%?

4. Roundtrip airline tickets to a traveler's destination of choice costs $378. What is the total price if the tax is 2.9%?

5. At Appliances-R-Us, they have a refrigerator for $596. What is the total price if the tax is 8.25%?

6. A child picks out a toy at the toy store that has a price tag of $18.85. What is the total price if the tax is 6%?

7. Jet skiing is a popular recreational activity but the equipment can be expensive. A boat dealership has one for $5,995. What is the total price if the tax is 4%?

8. When winter approaches, some people go out and buy a new coat. If a coat sells for $72 and the tax is 5%, how much is the total price?

9. At a sporting goods store, a man sees a pair of sunglasses that look just right. The sunglasses sell for $42. What is the total price if the tax is 6.5%?

10. At a beauty supply store, a woman finds some cosmetics that suit her fancy. If the cosmetics sell for $47.50 and the tax is 5.75%, how much is the total price?

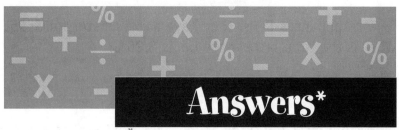

Answers*

*Answers to the preceding Practice Problems

1. Question: Golf is a serious game and so is the cost of the equipment. One popular set of golf clubs sells for $250. What is the total price if the sales tax is 4%?

> *Answer:*
> 4% is .04 in decimal form.
> The tax is .04 times 250 which equals 10.
> Add 250 to 10 to get the total price of $260.

2. Question: At Bright-Lights Electronics, they have a new television for $395. What is the total price if the sales tax is 7%?

> *Answer:*
> 7% is .07 in decimal form.
> The tax is .07 times 395 which equals 27.65.
> Add 395 to 27.65 to get the total price of $422.65.

3. Question: One night at a comfortable hotel costs a traveler $85.60 plus tax. What is the total price if the tax is 5.6%?

> *Answer:*
> 5.6% is .056 in decimal form.
> The tax is .056 times 85.60 which,
> rounded to the nearest hundredth, equals 4.79.
> Add 85.60 to 4.79 to get the total price of $90.39.

4. Question: Roundtrip airline tickets to a traveler's destination of choice costs $378. What is the total price if the tax is 2.9%?

> *Answer:*
> 2.9% is .029 in decimal form.
> The tax is .029 times 378 which,
> rounded to the nearest hundredth, equals 10.96.
> Add 378 to 10.96 to get the total price of $388.96.

5. Question: At Appliances-R-Us, they have a refrigerator for $596. What is the total price if the tax is 8.25%?

Answer:
8.25% is .0825 in decimal form.
The tax is .0825 times 596 which equals 49.17.
Add 596 to 49.17 to get the total price of $645.17.

6. Question: A child picks out a toy at the toy store that has a price tag of $18.85. What is the total price if the tax is 6%?

Answer:
6% is .06 in decimal form.
The tax is .06 times 18.85 which,
rounded off to the nearest hundredth, equals 1.13.
Add 18.85 to 1.13 to get the total price of $19.98.

7. Question: Jet skiing is a popular recreational activity but the equipment can be expensive. A boat dealership has one for $5,995. What is the total price if the tax is 4%?

Answer:
4% is .04 in decimal form.
The tax is .04 times 5,995 which equals 239.8.
Add 5,995 to 239.8 to get the total price of $6,234.80.

8. Question: When winter approaches, some people go out and buy a new coat. If a coat sells for $72 and the tax is 5%, how much is the total price?

Answer:
5% is .05 in decimal form.
The tax is .05 times 72 which equals 3.6.
Add 72 to 3.6 to get the total price of $75.60.

9. Question: At a sporting goods store, a man sees a pair of sunglasses that look just right. The sunglasses sell for $42. What is the total price if the tax is 6.5%?

> *Answer:*
> 6.5% is .065 in decimal form.
> The tax is .065 times 42 which equals 2.73.
> Add 42 to 2.73 to get the total price of $44.73.

10. Question: At a beauty supply store, a woman finds some cosmetics that suit her fancy. If the cosmetics sell for $47.50 and the tax is 5.75%, how much is the total price?

> *Answer:*
> 5.75% is .0575 in decimal form.
> The tax is .0575 times 47.5 which,
> rounded to the nearest hundredth, equals 2.73.
> Add 47.5 to 2.73 to get the total price of $50.23.

Chapter 11

Discounts

Everywhere you look there are advertisements for products announcing "Discount, Sale, Save, and Price cut." They all suggest that you will save money if you buy their product. These discounts are expressed in terms of a percentage or a fraction. If you are a seller, you need to figure out how much to charge your customer. If you are a buyer, you need to figure out how much it will cost you.

The first step in calculating the discount price of an item is to convert the discount expressed in terms of a percentage or a fraction into a decimal number. For a percentage, this is done by moving the decimal point two places to the left. Here are some examples.

5% = .05   10% = .10

15% = .15   20% = .20

25% = .25   40% = .40

50% = .50   75% = .75

In order to convert a discount expressed as a fraction into a decimal number, you need to divide. Divide the top number by the bottom number. Here are some examples.

$$\frac{1}{4} = 4\overline{)\begin{array}{r} .25 \\ 1.00 \end{array}}$$
$$\begin{array}{r} 80 \\ \hline 20 \\ 20 \\ \hline 0 \end{array}$$

$$\frac{1}{3} = 3\overline{)\begin{array}{r} .33 \\ 1.00 \end{array}}$$
$$\begin{array}{r} 90 \\ \hline 10 \\ 9 \\ \hline 1 \end{array}$$

$$\frac{1}{2} = 2\overline{)\begin{array}{r} .50 \\ 1.00 \end{array}}$$
$$\begin{array}{r} 1.00 \\ \hline 0 \end{array}$$

The second step is to multiply the original price of the item by the discount expressed in decimal form. This gives you the discount in dollar amount. Here is an example.

If a pair of pants regularly sells for $36.00, how much is the discount if it is being sold for 25% off?

$$\begin{array}{r} 36 \text{ original price} \\ \times\ .25 \text{ discount} \\ \hline 180 \\ 720 \\ \hline 9.00 \end{array}$$

➤ The discount is $9.00.

The third step is to subtract the discount from the original price.

$$\begin{array}{r} \$36.00 \text{ original price} \\ -\ 9.00 \text{ discount} \\ \hline \$27.00 \end{array}$$

➤ The discounted price is $27.00.

EXAMPLE WORD PROBLEM 1

A hardware store is having a sale, and they are offering a 30% discount on hammers. Hammers regularly sell for $24.00. How much is the discounted price?

Solution

30% in decimal form is .30.

Multiply $24.00 by .30 to find the discount.

> 24 original price
> x .30
> 7.20

 The discount is $7.20.

Subtract the discount from the original price.

> $24.00 original price
> - 7.20 discount
> $16.80

The discounted price is $16.80.

EXAMPLE WORD PROBLEM 2

Light bulbs come twenty-four to a case. A case regularly sells for $15.00. This week, light bulbs are on sale for 40% off. How much is the discounted price?

Solution

40% in decimal form is .40.

Multiply $15.00 by .40 to find the discount.

    15 original price
  x .40
   6.00   The discount is $6.00.

Subtract the discount from the original price.

  $15.00 original price
  - 6.00 discount
   $9.00

The discounted price is $9.00.

EXAMPLE WORD PROBLEM 3

A car dealer is having a year-end sale and all sports cars are 15% off. The one that you like has a price tag of $30,000. How much is the discounted price?

Solution

15% in decimal form is .15.

Multiply $30,000 by .15 to find the discount.

30,000 original price
x .15
4,500   The discount is $4,500.

Subtract the discount from the original price.

$30,000 orginal price
- 4,500 discount
$25,500

The discounted price is $25,500.

EXAMPLE WORD PROBLEM 4

At a pet store, all tortoises are 10% off. You are thinking about buying one that regularly costs $49.00. How much is the discounted price?

Solution

10% in decimal form is .10.

Multiply $49.00 by .10 to find the discount.

   49 original price
   x .10
   4.90   ▶▶   The discount is $4.90.

Subtract the discount from the original price.

   $49.00 orginal price
   - 4.90 discount
   $44.10

The discounted price is $44.10.

EXAMPLE WORD PROBLEM 5

A sporting goods store is selling tennis rackets at 1/3 off. If you wanted to buy a racket that regularly sells for $55.00, what would be the discounted price?

Solution

1/3 in decimal form is .33.

The decimal form .33 is obtained by dividing 1 by 3.

Multiply $55.00 by .33 to find the discount.

```
    55 original price
  x .33
  18.15
```
 The discount is $18.15.

Subtract the discount from the original price.

```
  $55.00 orginal price
  - 18.15 discount
  $36.85
```

The discounted price is $36.85.

EXAMPLE WORD PROBLEM 6

Blue Ribbon Bakery is having a sale on muffins. If you buy one dozen muffins at the regular price of $3.50, you get the second dozen for 1/2 off. How much are 2 dozen muffins?

Solution

1/2 in decimal form is .5.

Multiply $3.50 by .50 to find the discount.

    3.50 original price
    x .50
    1.75   The discount is $1.75.

Subtract the discount from the original price. Then, add the cost for each dozen together.

    $3.50 orginal price
    - 1.75 discount
     $1.75

> $3.50 cost for the first dozen
> + $1.75 cost for the second dozen
> $5.25 cost for two dozen muffins

EXAMPLE WORD PROBLEM 7

An antique dealer is having a sale. All antique clocks are 35% off. What would be the discounted price for a clock with a price tag of $240?

Solution

35% in decimal form is .35.

Multiply $240 by .35 to find the discount.

> 240 original price
> x .35
> 84.00

 The discount is $84.00.

Subtract the discount from the original price.

> $240 original price
> - 84 discount
> $156

The discounted price is $156.00.

EXAMPLE WORD PROBLEM 8

A popular champagne regularly sells for $42.00 a bottle. It is on sale now for 20% off. How much is the discounted price?

Solution

20% in decimal form is .20.

Multiply $42.00 by .20 to find the discount.

>   42 original price
> **x .20**
>   8.40   The discount is $8.40.

Subtract the discount from the original price.

> $42.00 original price
> **- 8.40** discount
> $33.60

The discounted price is $33.60.

EXAMPLE WORD PROBLEM 9

A storeowner needs a wall built. The best estimate that she could find is $2,500. Another contractor told her that he would beat her lowest estimate by 18%. How much will it cost to build the wall?

Solution

18% in decimal form is .18.

Multiply $2,500 by .18 to find the discount.

2500 original price
x .18
450.00   The discount is $450.

Subtract the discount from the original price.

$2,500 original price
- 450 discount
$2,050

The discounted price is $2,050.

EXAMPLE WORD PROBLEM 10

A boat dealer is offering a 7% discount on all boats in stock. What would be the discounted price for a boat that regularly sells for $32,000?

Solution

7% in decimal form is .07.

Multiply $32,000 by .07 to find the discount.

32,000 original price
x .07
2,240   The discount is $2,240.

Subtract the discount from the original price.

$32,000 original price
- 2,240 discount
$29,760

The discounted price is $29,760.

EXAMPLE WORD PROBLEM 11

At Acme Medical Insurance Company, you can receive a 7.5% discount for being a non-smoker. If you were a non-smoker and your medical insurance policy normally cost $1,240 a year, what would be the discounted price?

Solution

7.5% in decimal form is .075.

Multiply $1,240 by .075 to find the discount.

    1,240 original price
    x .075
      93.0   The discount is $93.

Subtract the discount from the original price.

    $1,240 original price
    - 93 discount
    $1,147

The discounted price is $1,147.

EXAMPLE WORD PROBLEM 12

At C-U-World Travel Agency, they give a 20.25% discount on all flights overseas during the off season. The destination of your choice regularly costs $1,420. What is the discounted price?

Solution

20.25% in decimal form is .2025.

Multiply $1,420 by .2025 to find the discount.

```
    1,420 original price
  x .2025
   287.55
```

  The discount is $287.55.

Subtract the discount from the original price.

```
  $1,420.00 original price
   - 287.55 discount
  $1,132.45
```

The discounted price is $1,132.45.

Practice Problems*

*Answers to Practice Problems on Pages 171-175

1. An industrial supply company gives non-profit organizations a 12% discount. How much would a $1,500 order cost after discount?

2. A restaurant gives senior citizens a 20% discount on meals. How much would an $8.50 meal cost a senior citizen?

3. A business conference gives a 5% discount to all registrants who pay in advance. The regular price is $320. How much is the discounted price?

4. Downtown Clothing Store sells dress shirts for $36.00. If you buy two shirts, you get the second one for 1/4 off. How much do two shirts cost?

5. Save-A-Lot Furniture Store is selling all sofas for 1/3 off. The one that you like has a price tag of $500. How much is the discounted price?

6. Uptown Music Store is going out of business. All CDs are on sale for 70% off. A CD regularly sells for $20.00. How much is the discounted price?

7. We-Got-It department store gives its employees a 10% discount on all purchases. If an employee wants to buy a refrigerator that regularly sells for $750, how much will the discounted price be?

8. A computer manufacturing company gives its largest customer a 23.75% discount on all orders. This customer placed an order with a regular price of $25,000. What is the discounted price?

9. Plug-It-In Appliances is having a sale. All washers and dryers are 17% off. What is the discounted price for a washer that regularly sells for $475?

10. A carpet store is having a 45% off sale. A homeowner estimates that it would cost her $2,700 to carpet her whole house before the discount. How much is the discounted price?

Answers*

*Answers to the preceding Practice Problems

1. Question: An industrial supply company gives non-profit organizations a 12% discount. How much would a $1,500 order cost after discount?

> *Answer:*
> 12% is .12 in decimal form.
> The discount is .12 times 1,500 which equals 180.
> Subtract $180 from $1,500 to get
> the discounted price of $1,320.

2. Question: A restaurant gives senior citizens a 20% discount on meals. How much would an $8.50 meal cost a senior citizen?

> *Answer:*
> 20% is .20 in decimal form.
> The discount is .20 times 8.5 which equals 1.70.
> Subtract $1.70 from $8.50 to get
> the discounted price of $6.80.

3. Question: A business conference gives a 5% discount to all registrants who pay in advance. The regular price is $320. How much is the discounted price?

Answer:
5% is .05 in decimal form.
The discount is .05 times 320 which equals 16.
Subtract $16 from $320 to get
the discounted price of $304.

4. Question: Downtown Clothing Store sells dress shirts for $36.00. If you buy two shirts, you get the second one for 1/4 off. How much do two shirts cost?

Answer:
1/4 is .25 in decimal form.
The discount is .25 times 36 which equals 9.
The cost of the second shirt is $36.00 minus $9.00,
which equals $27.00. The cost for two shirts is
$36.00 plus $27.00, which equals $63.00.

5. Question: Save-A-Lot Furniture Store is selling all sofas for 1/3 off. The one that you like has a price tag of $500. How much is the discounted price?

Answer:
1/3 is .33 in decimal form.
The discount is .33 times 500 which equals 165.
Subtract $165 from $500 to get
the discounted price of $335.

6. Question: Uptown Music Store is going out of business. All CDs are on sale for 70% off. A CD regularly sells for $20.00. How much is the discounted price?

Answer:
70% is .70 in decimal form.
The discount is .70 times 20 which equals 14.
Subtract $14.00 from $20.00 to get
the discounted price of $6.00.

7. Question: We-Got-It department store gives its employees a 10% discount on all purchases. If an employee wants to buy a refrigerator that regularly sells for $750, how much will the discounted price be?

> *Answer:*
> 10% is .10 in decimal form.
> The discount is .10 times 750 which equals 75.
> Subtract $75 from $750 to get
> the discounted price of $675.

8. Question: A computer manufacturing company gives it largest customer a 23.75% discount on all orders. This customer placed an order with a regular price of $25,000. What is the discounted price?

> *Answer:*
> 23.75% is .2375 in decimal form.
> The discount is .2375 times 25,000 which equals 5,937.50.
> Subtract $5,937.50 from $25,000.00 to get
> the discounted price of $19,062.50.

9. Question: Plug-It-In Appliances is having a sale. All washers and dryers are 17% off. What is the discounted price for a washer that regularly sells for $475?

> *Answer:*
> 17% is .17 in decimal form.
> The discount is .17 times 475 which equals 80.75.
> Subtract $80.75 from $475.00 to get
> the discounted price of $394.25.

10. Question: A carpet store is having a 45% off sale. A homeowner estimates that it would cost her $2,700 to carpet her whole house before the discount. How much is the discounted price?

> *Answer:*
> 45% is .45 in decimal form.
> The discount is .45 times 2,700 which equals 1,215.
> Subtract $1,215 from $2,700 to get
> the discounted price of $1,485.

Chapter 12

Measurements

Measurements are numbers that describe the amount or quantity of some attribute. These attributes can be time, distance, speed, weight, area, volume, and temperature. Further, these attributes come in different units of measurement.

Time is an essential measurement whether it is being expressed in terms of seconds, minutes, hours, days, weeks, months, or years. When it comes to keeping time, there is standard time and there is military time. Distance can also be expressed in various units of measurement, including inches, feet, yards, and miles. Distance is a one-dimensional measurement. Speed equals distance divided by time. Speed, distance, and time are all related. If you know any two of them, you can figure out the third.

Weight is a common measurement. It can be expressed in grams, ounces, pounds, and tons. Area is an important measurement that can be expressed in terms of square inches, square feet, and square yards. In addition, area can be measured in square miles and acres. Area is a two-dimensional measurement. Volume is another measurement. For solids, volume can be measured in terms of cubic inches, cubic feet, or cubic yards. For liquids, volume can be measured in ounces, cups, pints, quarts, and gallons. Volume is a three-dimensional measurement. Temperature can be measured in degrees Fahrenheit or degrees centigrade. Here are some examples of measurements.

 ## TIME

1 Minute = 60 seconds
1 Hour = 60 minutes
1 Day = 24 hours
1 Month can equal anywhere from 28 to 31 days.
1 Year = 365 days
But every fourth year, called a leap year, has 366 days.

Military time is different from standard time. Military time uses a 24-hour cycle and does not use a.m. or p.m. In addition to the military, law enforcement and hospitals use military time. For example, 3:00 p.m. in standard time is 1500 in military time. It is read "fifteen hundred hours," but it is really 15 hours and 0 minutes after midnight. The minute after 1459 is 1500 because 60 minutes makes an hour just like in standard time.

Standard Time	Military Time	Standard Time	Military Time
Midnight	0000	Noon	1200
1:00 a.m.	0100	1:00 p.m.	1300
2:00 a.m.	0200	2:00 p.m.	1400
3:00 a.m.	0300	3:00 p.m.	1500
4:00 a.m.	0400	4:00 p.m.	1600
5:00 a.m.	0500	5:00 p.m.	1700
6:00 a.m.	0600	6:00 p.m.	1800
7:00 a.m.	0700	7:00 p.m.	1900
8:00 a.m.	0800	8:00 p.m.	2000
9:00 a.m.	0900	9:00 p.m.	2100
10:00 a.m.	1000	10:00 p.m.	2200
11:00 a.m.	1100	11:00 p.m.	2300

 EXAMPLE WORD PROBLEM 1

A surgery took a team of doctors 4 hours and 38 minutes to complete. How many minutes did the surgery take?

Solution

4 x 60 = 240 minutes

240 + 38 = 278 minutes

EXAMPLE WORD PROBLEM 2

If the military time is 18:30, what is the standard time?

Solution

18:00 is 6:00 p.m., so 18:30 is 6:30 p.m.

 ## DISTANCE

1 Foot = 12 inches
1 Yard = 3 feet
1 Yard = 36 inches
1 Mile = 5,280 feet

 ## EXAMPLE WORD PROBLEM 1

If there are 5,280 feet in one mile, how many yards are there in 4 miles?

Solution

First, divide the number of feet in a mile by 3 in order to find the number of yards in a mile.

5,280
<u>÷ 3</u>
1,760 yards.

Next, multiply 1,760 by 4 in order to find the number of yards in 4 miles.

1,760
<u>x 4</u>
7,040

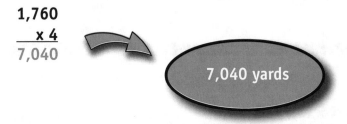

7,040 yards

EXAMPLE WORD PROBLEM 2

A carpenter needs three pieces of two by four wood for a project. The first piece needs to be 2 feet 8 inches. The second piece needs to be 1 foot 6 inches. The third piece needs to be 3 feet 5 inches. If these three pieces are to be cut from one board, how long should the board be?

Solution

2'8"
1'6"
<u>+ 3'5"</u>
6'19"

Then, simplify.

6'19" = 7'7"

EXAMPLE WORD PROBLEM 3

Planning a wedding involves some calculations. A wedding planner has 8 tables that will be used for a reception. Each table is 72 inches long. If the tables were arranged end to end, how long would they be?

Solution

One way to solve this problem would be to convert 72 inches to feet by dividing by 12 (6 feet). Then, multiply the number of tables (8) by 6 to get 48 feet. Another way to solve this problem is to multiply 72 times 8 to get 576 inches. Then, divide 576 by 12 to get 48 feet.

 ## EXAMPLE WORD PROBLEM 4

A carpenter is building a fence that runs from the edge of a house and is 368 feet long. There needs to be a post every 8 feet. How many posts does the builder need?

Solution

368
÷ 8
46

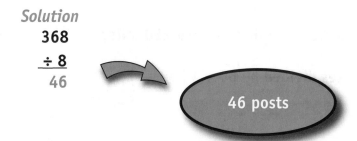

46 posts

SPEED

$$\text{Speed} = \frac{\text{Distance}}{\text{Time}}$$

Distance = Speed x Time

$$\text{Time} = \frac{\text{Distance}}{\text{Speed}}$$

Speed equals distance divided by time. Speed, distance, and time are all related. If you know any two of them, you can figure out the third.

 EXAMPLE WORD PROBLEM 1

If you are traveling in a car at 60 miles per hour, you are traveling at the rate of 1 mile per minute. At this rate, how long will it take to travel 250 miles?

Solution

It will take 250 minutes to drive 250 miles.

250 minutes divided by 60 = 4 hours and 10 minutes.

EXAMPLE WORD PROBLEM 2

A train travels 141 miles in 3 hours. What was its average rate of speed?

Solution

$$\text{Speed} = \frac{\text{Distance}}{\text{Time}}$$

$$\text{Speed} = \frac{141 \text{ miles}}{3 \text{ hours}}$$

Speed = 47 MPH

 EXAMPLE WORD PROBLEM 3

A person walked at the rate of 3.5 miles per hour for a total of 2.5 hours. What is the distance that this person traveled?

Solution
Distance = Speed x Time

Distance = 3.5 x 2.5

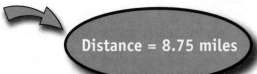

Distance = 8.75 miles

EXAMPLE WORD PROBLEM 4

A jet airplane traveled 1,870 miles at the rate of 440 miles per hour. How long did this trip take?

Solution

$$\text{Time} = \frac{\text{Distance}}{\text{Speed}}$$

$$\text{Time} = \frac{1{,}870 \text{ miles}}{440 \text{ MPH}}$$

Time = 4.25 hours

 WEIGHT

1 Ounce = 28.3 grams
1 Pound = 16 ounces
1 Ton = 2,000 pounds

 EXAMPLE WORD PROBLEM 1

A jewelry dealer purchases three gold chains weighing 17.8 grams, 35 grams, and 27.5 grams. In terms of ounces and grams, how much do the three gold chains weigh together?

Solution

```
    17.8
    35
  + 27.5
    80.3 grams
```

Next, find out how many ounces this is.

```
    80.3 total grams
  ÷ 28.3 grams per ounce
    2    ounces plus a remainder
```

2 ounces = 2 x 28.3 = 56.6 grams

```
    80.3
  - 56.6
    23.7 grams
```

80.3 grams = 2 ounces and 23.7 grams

 EXAMPLE WORD PROBLEM 2

How many pounds does 25 tons equal?

Solution

```
    2000
   x 25
  50,000
```

50,000 pounds

EXAMPLE WORD PROBLEM 3

Candle-Wick is a company that manufactures candles. Their standard candle weighs 5 ounces. In order to calculate shipping charges for a case of candles, a worker needs to know the weight of 24 candles.

Solution

24 candles
x 5 ounces per candle
120 ounces

120 ounces
÷ 16 ounces per pound
7.5 pounds

7 pounds, 8 ounces

 # EXAMPLE WORD PROBLEM 4

How many tons does 64,800 pounds equal?

Solution

64,800
÷ 2,000
32.4

32.4 tons

AREA

1 Square Foot = 144 square inches
1 Square Yard = 9 square feet
1 Acre = 43,560 square feet
1 Square Mile = 640 acres

 ## EXAMPLE WORD PROBLEM 1

How many square inches are there in a standard sheet of paper that measures 8 1/2 inches by 11 inches?

Solution

```
  11 length
x 8.5 width
  93.5
```

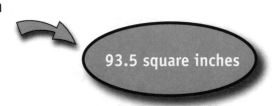

93.5 square inches

 ## EXAMPLE WORD PROBLEM 2

A rectangle shaped house is 40 feet in length and 37 feet wide. How many square feet is this house?

Solution

```
  40 length
x 37 width
1,480
```

1,480 square feet

EXAMPLE WORD PROBLEM 3

How many square yards are there in 162 square feet?

Solution

162 square feet

÷ 9 square feet per square yard

18

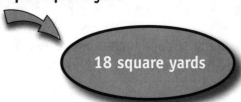

18 square yards

 EXAMPLE WORD PROBLEM 4

A ranch in Big Sky Country covers an area of 3,200 acres. How many square miles is this?

Solution

3,200 acres

÷ 640 acres per square mile

5

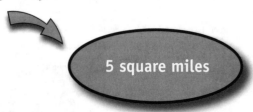

5 square miles

VOLUME

SOLIDS:
1 Cubic Foot = 1,728 cubic inches
1 Cubic Yard = 27 cubic feet

LIQUIDS:
1 Cup = 8 fluid ounces
1 Pint = 2 cups
1 Quart = 4 cups
1 Gallon = 4 quarts

 ## EXAMPLE WORD PROBLEM 1

On-The-Level Construction is working on one project that requires the pouring of a large cement patio in the backyard of a house. The architect figures that it will take 1,026 cubic feet of concrete. How many cubic yards is this?

Solution
1,026 cubic feet
÷ 27 cubic feet per cubic yard
      38

38 cubic yards

EXAMPLE WORD PROBLEM 2

During a kitchen-remodeling project, a couple decides to visit Ok-Appliances to pick out a new refrigerator. They finally decide on one that they both like, which can hold 21.8 cubic feet. How many cubic inches can this refrigerator hold?

Solution

1,728 cubic inches per cubic foot
x 21.8 cubic feet
37,670.4

37,670.4 cubic inches

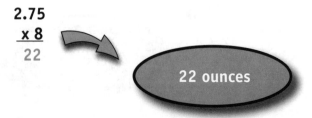 **EXAMPLE WORD PROBLEM 3**

How many ounces are there in 2 and 3/4 cups?

Solution
2 and 3/4 in decimal form is 2.75.
There are 8 ounces in a cup.

2.75
x 8
22

22 ounces

EXAMPLE WORD PROBLEM 4

How many gallons are there in 246 quarts?

Solution
There are 4 quarts in one gallon.

246
÷ 4
61.5

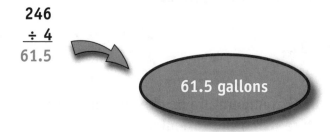

61.5 gallons

 ## TEMPERATURE

There are two commonly used temperature scales. The Fahrenheit temperature scale is widely used in the United States, but the Celsius temperature scale is equally useful. The Celsius scale is also called the Centigrade scale.

Using the Fahrenheit temperature scale, water boils at 212 degrees Fahrenheit. Using the Celsius temperature scale, water boils at 100 degrees Celsius.

Using the Fahrenheit temperature scale, water freezes

at 32 degrees Fahrenheit. Using the Celsius temperature scale, water freezes at 0 degrees Celsius.

At –40 degrees, Fahrenheit equals Celsius.

A Celsius temperature can be converted to a Fahrenheit temperature using the formula:

$$F = \frac{9}{5}C + 32$$

A Fahrenheit temperature can be converted to a Celsius temperature using the formula:

$$C = \frac{5}{9}(F - 32)$$

 EXAMPLE WORD PROBLEM 1

A traveler from the United States is visiting another country where they use the Celsius temperature scale. The temperature on the thermometer reads 25 degrees Celsius. How many degrees Fahrenheit is this?

Solution

$$F = \frac{9}{5}(25) + 32 \qquad F = \frac{225}{5} + 32$$

```
   45
 + 32
   77
```

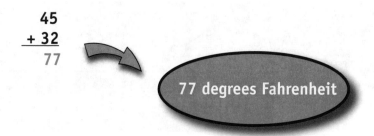

77 degrees Fahrenheit

EXAMPLE WORD PROBLEM 2

A traveler visiting the United States comes from a foreign country that uses the Celsius temperature scale. The temperature on the thermometer reads 86 degrees Fahrenheit. How many degrees Celsius is this?

Solution

$$C = \frac{5}{9}(86 - 32) \qquad C = \frac{5}{9}(54)$$

$$\frac{270}{9} = 30$$

30 degrees Celsius

*Answers to Practice Problems on Pages 196-200

1. A boat captain tells his crew that there will be an assembly at 1600 hours. At what time is the assembly in standard time?

2. The area of a front lawn is 85 square yards. How many square feet is this lawn?

3. A car traveled 100 miles in 2 and 1/2 hours. What was its average rate of speed?

4. How many ounces are there in 4 and 3/4 pounds?

5. A property owner has a piece of land that is 1/4 acre big. How many square feet is this?

6. A parent is planning a birthday party for a child. The parent has 3 and 1/4 gallons of fruit punch. How many cups is this?

7. A thermometer reads 20 degrees Celsius. How many degrees Fahrenheit is this?

8. A train traveled at the rate of 30 miles per hour for 5 and 1/2 hours. How far did the train travel?

9. How many cubic feet are there in 20 cubic yards?

10. The temperature on a Fahrenheit thermometer reads 59 degrees. How many degrees Celsius is this?

Answers*

*Answers to the preceding Practice Problems

1. Question: A boat captain tells his crew that there will be an assembly at 1600 hours. At what time is the assembly in standard time?

> *Answer:*
> 1600 is 16 hours after midnight or 4:00 p.m.

2. Question: The area of a front lawn is 85 square yards. How many square feet is this lawn?

> *Answer:*
> There are 9 square feet in one square yard.
> 85 x 9 = 765 square feet

3. Question: A car traveled 100 miles in 2 and 1/2 hours. What was its average rate of speed?

> ### Answer:
> **100 ÷ 2.5 = 40 miles per hour**

4. Question: How many ounces are there in 4 and 3/4 pounds?

> ### Answer:
> **There are 16 ounces in a pound.**
> **4.75 x 16 = 76 ounces**

5. Question: A property owner has a piece of land that is 1/4 acre big. How many square feet is this?

> ### Answer:
> **There are 43,560 square feet per acre.**
> **43,560 ÷ 4 = 10,890 square feet**

6. **Question:** A parent is planning a birthday party for a child. The parent has 3 and 1/4 gallons of fruit punch. How many cups is this?

Answer:
There are 4 quarts in a gallon and 4 cups in a quart. Therefore, there are 16 cups in a gallon.
3.25 x 16 = 52 cups

7. **Question:** A thermometer reads 20 degrees Celsius. How many degrees Fahrenheit is this?

Answer:

$$F = \frac{9}{5}C + 32 \quad \text{formula}$$

$$F = \frac{9}{5}(20) + 32 \qquad F = \frac{180}{5} + 32$$

36 + 32 = 68 degrees Fahrenheit

8. Question: A train traveled at the rate of 30 miles per hour for 5 and 1/2 hours. How far did the train travel?

Answer:
30 x 5.5 = 165 miles

9. Question: How many cubic feet are there in 20 cubic yards?

Answer:
There are 27 cubic feet in one cubic yard.
20 x 27 = 540 cubic feet

10. Question: The temperature on a Fahrenheit thermometer reads 59 degrees. How many degrees Celsius is this?

Answer:

$$C = \frac{5}{9}(F - 32) \quad \text{formula}$$

$$C = \frac{5}{9}(59 - 32) \qquad C = \frac{5}{9}(27)$$

$$\frac{135}{9} = 15 \text{ degrees Celsius}$$

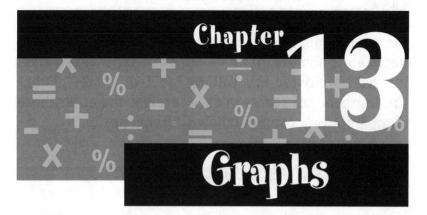

Chapter 13

Graphs

A graph is a picture of a set of numbers. A graph displays a set of numbers so that any patterns can be seen. Graphs allow you to communicate information in an understandable manner. It has been said that, "A picture is worth a thousand words." Likewise, "A graph is worth a thousand numbers."

Sometimes, before a set of numbers is graphed, it is put into a special format called a "frequency distribution." A frequency distribution is two columns of numbers. The first column has scores in it and the second column has frequencies. Graphing a frequency distribution of scores allows you to see how frequently particular scores or groups of scores occurred relative to others.

The exact method for creating a frequency distribution depends on the range of the scores. If the scores have a narrow range of values, then list this range from the minimum score to the maximum score. Put this list in a column labeled "scores". Then, count the frequency of each score and enter these numbers into a column labeled "frequency". The other possibility occurs when the scores have a wide range of values. In this situation, the scores must be broken down into class intervals. Between 8 and 20 intervals are sufficient. These class intervals are listed in a column labeled "scores". Count the frequency of scores in each interval and enter these numbers into a column labeled "frequency".

 FREQUENCY HISTOGRAM

A frequency histogram is a graph of a frequency distribution. A frequency distribution breaks a set of numbers up into intervals and counts how many numbers fall into each interval.

 EXAMPLE 1

Given a set of 20 scores, create a frequency distribution. Then, using this frequency distribution, draw a histogram. Since there is a small range of scores, the frequency of individual scores can be counted.

Scores: 7, 4, 8, 7, 9, 5, 4, 8, 6, 2, 2, 9, 5, 7, 4, 5, 7, 3, 6, 4

Frequency Distribution:

Score	Frequency
9	2
8	2
7	4
6	2
5	3
4	4
3	1
2	2
	20 scores

Histogram

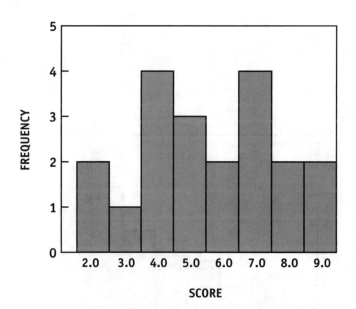

EXAMPLE 2

Given a frequency distribution that contains 100 scores, draw a histogram. Since the range of these scores is large, class intervals were used to divide the score up into groups. Each class interval is 5 scores wide. In addition to the frequency distribution, which only includes the class intervals and the frequency counts, the midpoint of each class interval is listed. These midpoints can be seen on the axis labeled "Score" on the histogram.

Frequency Distribution:

Class Interval	Frequency	Midpoint
96 – 100	25	98
91 – 95	28	93
86 – 90	14	88
81 – 85	11	83
76 – 80	9	78
71 – 75	5	73
66 – 70	7	68
61 – 65	1	63

Histogram

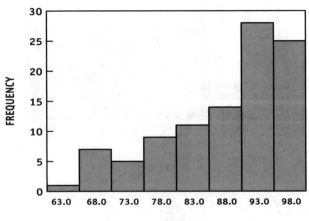

EXAMPLE 3

The frequency histogram below represents 500 scores. The average for this set of scores is 50.

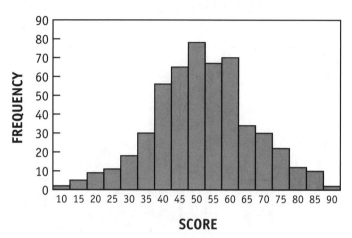

 EXAMPLE 4

The frequency histogram below represents 1000 scores. The average for this set of scores is 100.

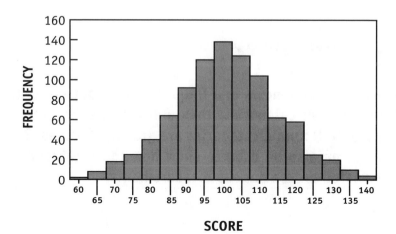

EXAMPLE 5

The frequency histogram below represents 30 scores. The average for this set of scores is 5.2.

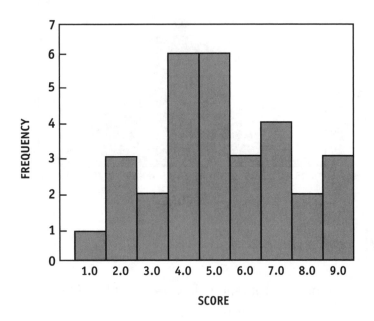

 BAR GRAPH

When the frequencies in a frequency distribution come from nominal categories, they are best displayed in a bar graph. Nominal categories are groups of scores. Bar graphs can also be used to show how these categories are related on continuous measurements.

EXAMPLE 1

In a metropolitan city, a sample of 150 people were surveyed. Participants were asked, "What is your main mode of transportation?" The results from this survey are displayed in the form of a frequency distribution and a bar graph.

Frequency Distribution:

Category	Frequency
Car	90
Bus	10
Subway	25
Taxi	5
Walk	3
Bicycle	2
Train	15
	150 people surveyed

Bar Graph

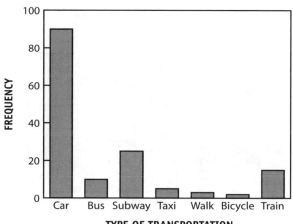

EXAMPLE 2

In a survey, 200 elementary school children were asked, "What is your favorite subject in school?" The results from this survey are displayed in the form of a frequency distribution and a bar graph.

Frequency Distribution:

Category	Frequency
English	30
History	24
Math	36
Science	32
Art	44
P.E.	34
	200 children surveyed

Bar Graph

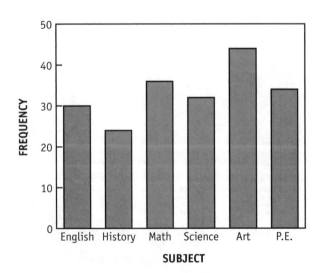

EXAMPLE 3

Companies are often interested in comparing their sales figures to that of their competitors. In one area of the country, five different supply companies control the sales of wood and hardware. The total sales last year for these companies are listed followed by a bar graph.

Company	Sales Last Year
Company A	$1,200,000
Company B	$4,400,000
Company C	$3,200,000
Company D	$3,200,000
Company E	$2,800,000

Bar Graph

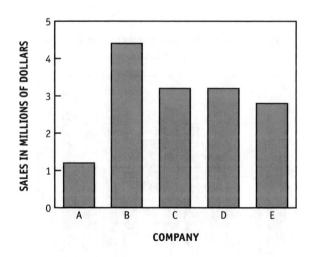

EXAMPLE 4

Greenville gets a good bit of rain. The average inches of rainfall for each month is displayed in the table below followed by a bar graph.

Month	Rainfall (inches)
January	7
February	4
March	5
April	7
May	9
June	10
July	8
August	6
September	9
October	10
November	11
December	10

Bar Graph

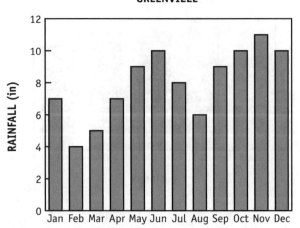

AVERAGE MONTHLY RAINFALL FOR THE CITY OF GREENVILLE

PIE CHART

While the frequencies for a set of nominal categories is displayed with a bar graph, the *relative* frequencies are displayed using a pie chart. In essence, a pie chart compares parts of a whole. In a pie chart, all of the percentages add up to 100%. Here are some examples of pie charts.

EXAMPLE 1

The enrollment figures at one public college reveals that 52% of the student body are female while 48% are male.

Pie Chart

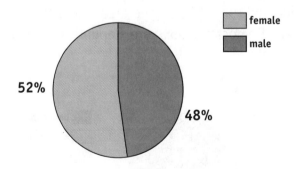

EXAMPLE 2

Outside a movie theater, an exit survey taker asked the moviegoers, "Would you recommend this movie to a friend?" 62% said, "yes," while 38% said, "no."

Pie Chart

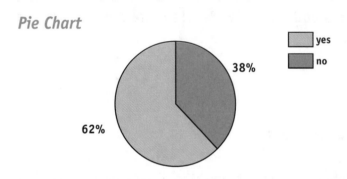

 EXAMPLE 3

A political poll of voters in one area found that 48% of the voters are Democrats, 35% are Republicans, and 17% are Independents.

Pie Chart

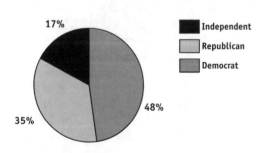

EXAMPLE 4

A small company has its financial assets in four different investments.

Investment	Percent of Assets
Stocks	20%
Bonds	20%
Cash	45%
Mutual Funds	15%

Pie Chart

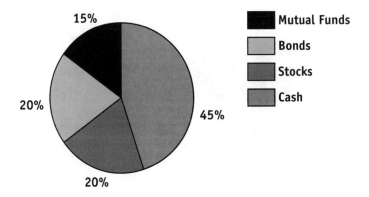

EXAMPLE 5

The percentage of students earning each grade in a high school geometry class is listed below followed by a pie chart.

Grade	Percent of Class
A	20%
B	25%
C	30%
D	15%
F	10%

Pie Chart

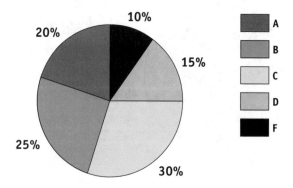

EXAMPLE 6

Residents of one city were surveyed to collect demographic information. The results from this survey give a breakdown according to race.

Race	Percent of City
White	31%
Black	14%
Asian	23%
Hispanic	27%
Other	5%

Pie Chart

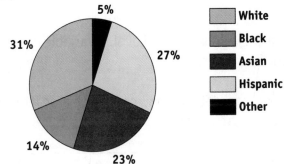

 ## SCATTER PLOT

A scatter-plot is a graph that shows the relationship between two variables. A scatter-plot is just a set of (x, y) coordinates. Each coordinate is a point on the graph. The x value represents a measurement on one variable, while the y value represents a measurement on another variable. Remember that a variable is just a measurement that can take on more than one value. A scatter-plot is useful because in one picture you can see if there is a relationship between two variables.

 EXAMPLE 1

A research study was conducted to see if there was any relationship between grade level in school and the amount of homework each week. A sample of 12 students was surveyed.

Given: Variable x is Grade Level. Variable y is Hours of Homework Each Week.

Task: Using the data below, construct a scatter plot and describe the relationship.

x	Grade Level	1	2	3	4	5	6	7	8	9	10	11	12
y	Hours of Homework	2	3	3	6	4	10	7	10	12	9	14	15

Solution

You can think of these two variables as one set of (x, y) coordinates on a graph. Remember that a coordinate is just a point. Graph the following points: (1, 2), (2, 3), (3, 3), (4, 6), (5, 4), (6, 10),

(7, 7), (8, 10), (9, 12), (10, 9), (11, 14), & (12, 15). This graph shows that as grade level increases, the number of homework hours tends to increase.

Scatter Plot

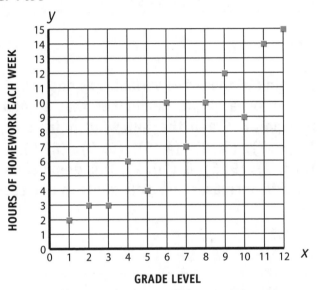

EXAMPLE 2

A research study was conducted to see if there was any relationship between math skill and art skill. A sample of 12 people agreed to participate in the study. Each person took two tests. One was an art test. The other was a math test.

Given: Variable x is a math test score. Variable y is an art test score.

Basic Math Refresher

Task: Using the data below, construct a scatter plot and describe the relationship between the two variables.

x	Math Test		1	2	3	3	4	5	6	7	7	8	9	10
y	Art Test		11	9	7	9	6	9	6	7	4	3	5	2

Solution

You can think of these two variables as one set of (x, y) coordinates on a graph. Remember that a coordinate is just a point. Graph the following points: (1, 11), (2, 9), (3, 7), (3, 9), (4, 6), (5, 9), (6, 6), (7, 7), (7, 4), (8, 3), (9, 5), & (10, 2). This graph shows that as math skill increases, art skill tends to decrease.

Scatter Plot

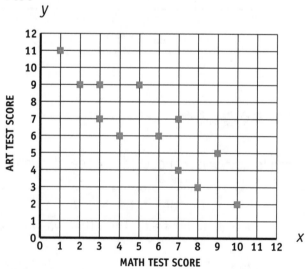

CORRELATION

A statistical technique called correlation tells you if two measurements are related along a straight line. A scatter-plot is one way of looking at correlation. There are three different types of correlation.

 ## POSITIVE CORRELATION

In a positive correlation, as one measurement increases, the other measurement also increases.

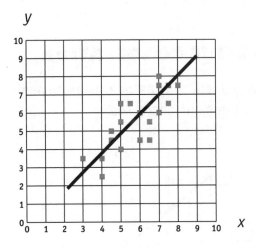

ZERO CORRELATION

In a zero correlation, the two measurements are not related to each other along a straight line.

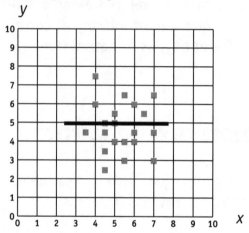

NEGATIVE CORRELATION

In a negative correlation, as one measurement increases, the other measurement decreases.

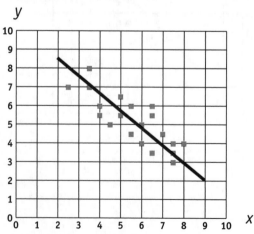

LINE GRAPH

A line graph is a way of showing how two measurements are related. A line can also be used to show how something changes over time. When creating a line graph where time is one of the measurements, time should be displayed on the x-axis, which is the horizontal axis. Here are some examples.

 ## EXAMPLE 1

The average temperature for the city of Appleton was recorded for five consecutive days. Using these temperatures, a line graph was created to show changes in temperature over time.

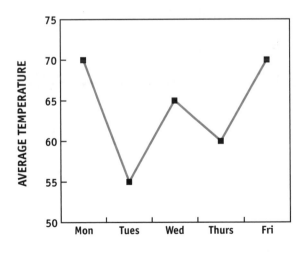

EXAMPLE 2

The graph below shows the interest rate paid by two different banks on a savings account. The interest rate depends upon the amount of the balance.

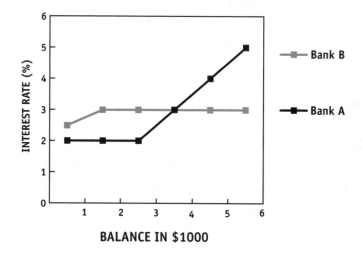

Practice Problems*

*Answers to Practice Problems on Pages 227-229

1. A classroom of thirty high school students took a pop quiz in math class. A frequency histogram of the students' scores is shown below. How many students earned a score of 5? What is the most frequently occurring score?

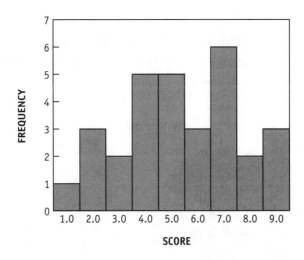

2. The average monthly rainfall for the city of Pinewood is shown in a bar graph below. Which month had the third highest amount of rainfall? How many inches of rain did Pinewood receive that month?

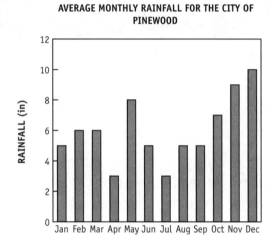

AVERAGE MONTHLY RAINFALL FOR THE CITY OF PINEWOOD

3. A group of sports fans was asked, "What is your favorite sport?" The responses are displayed in the pie chart below. What percent of the people gave an answer other than football?

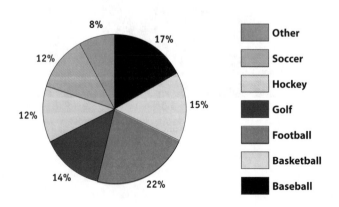

4. A scatter plot showing the relationship between two measurements is shown below. Judging by the pattern of the points, is this a positive correlation, a zero correlation, or a negative correlation?

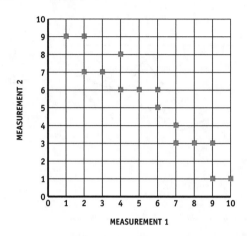

MEASUREMENT 1

5. The annual sales for a pineapple grower were recorded for an eight-year period and a line graph was created. How much did this producer sell in 2003?

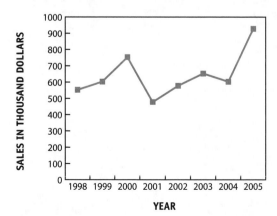

YEAR

Basic Math Refresher

6. At a car dealership, the number of cars sold by two different salespeople for a six-month period is displayed in the line graph below. During the months of May and June, how many more cars did salesperson A sell than salesperson B?

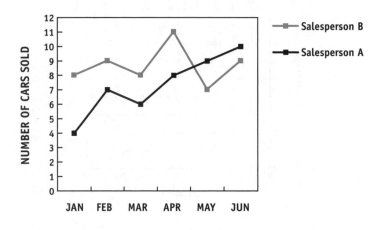

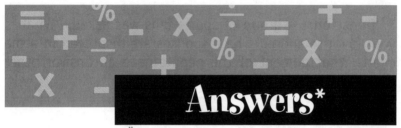

Answers*

*Answers to the preceding Practice Problems

1. Question: A classroom of thirty high school students took a pop quiz in math class. A frequency histogram of the students' scores is shown. How many students earned a score of 5? What is the most frequently occurring score?

Answer:
Five students earned a score of 5.
The most frequently occurring score is 7.

2. Question: The average monthly rainfall for the city of Pinewood is shown in a bar graph. Which month had the third highest amount of rainfall? How many inches of rain did Pinewood receive that month?

Answer:
The month of May had the third highest amount of rainfall. May delivered 8 inches of rain to Pinewood.

3. **Question:** A group of sports fans was asked, "What is your favorite sport?" The responses are displayed in a pie chart. What percent of the people gave an answer other than football?

Answer:
78% of the people gave a favorite sport other than football. There are two different ways to arrive at this answer. One way is by adding up the percentages for all of the sports other than football.
A simpler way is to take 100% - 22% = 78%.
This is true because all of the categories have to add up to 100%. After subtracting football, what you have left is everything else.

4. **Question:** A scatter plot showing the relationship between two measurements is shown. Judging by the pattern of the points, is this a positive correlation, a zero correlation, or a negative correlation?

Answer:
This is a scatter plot of a negative correlation.
In a negative correlation, as one measurement increases, the other measurement decreases.

5. Question: The annual sales for a pineapple grower were recorded for an eight-year period and a line graph was created. How much did this producer sell in 2003?

Answer:
In 2003, this producer sold $650,000 worth of pineapples.

6. Question: At a car dealership, the number of cars sold by two different salespeople for a six-month period is displayed in the line graph below. During the months of May and June, how many more cars did salesperson A sell than salesperson B?

Answer:
Salesperson A outsold Salesperson B by 2 cars in May and 1 car in June to make the final answer 3 cars.